Lecture Notes in Computer Scie

Edited by G. Goos, J. Hartmanis and J. van Le

T0250726

Advisory Board: W. Brauer D. Gries J. Stoer

Springer
Berlin
Heidelberg
New York
Barcelona
Budapest
Hong Kong
London
Milan
Paris
Santa Clara
Singapore
Tokyo

Rangachar Kasturi   Karl Tombre (Eds.)

Graphics Recognition

Methods and Applications

First International Workshop
University Park, PA, USA, August 10-11, 1995
Selected Papers

 Springer

Series Editors

Gerhard Goos, Karlsruhe University, Germany

Juris Hartmanis, Cornell University, NY, USA

Jan van Leeuwen, Utrecht University, The Netherlands

Volume Editors

Rangachar Kasturi
Pennsylvania State University
University Park, PA 16802-6106, USA

Karl Tombre
INRIA Lorraine & CRIN/CNRS
F-54602 Villers les Nancy Cedex, France

Cataloging-in-Publication data applied for

Die Deutsche Bibliothek - CIP-Einheitsaufnahme

Graphics recognition : methods and applications ; first
international workshop, University Park, PA, USA, August 10 -
11, 1995 ; selected papers / Rangachar Kasturi ; Karl Tombre
(ed.). - Berlin ; Heidelberg ; New York ; Barcelona ; Budapest ;
Hong Kong ; London ; Milan ; Paris ; Santa Clara ; Singapore ;
Tokyo : Springer, 1996
  (Lecture notes in computer science ; Vol. 1072)
  ISBN 3-540-61226-2
NE: Kasturi, Rangachar [Hrsg.]; GT

CR Subject Classification (1991): I.5, I.4, I.2.1, I.2.8, G.2.2,F.2.2

ISBN 3-540-61226-2 Springer-Verlag Berlin Heidelberg New York

© Springer-Verlag Berlin Heidelberg 1996
Printed in Germany

Typesetting: Camera-ready by author
SPIN 10512889   06/3142 – 5 4 3 2 1 0   Printed on acid-free paper

Preface

Document image analysis is an exciting field of pattern recognition, in which much research activity is taking place, not only on optical character recognition (which is probably the first topic to come to mind when somebody mentions document analysis), but also on the analysis of physical and logical structures of scanned documents, and more generally the interpretation of the document's contents. One of the topical fields is *graphics recognition*, which deals with problems such as raster-to-vector techniques, recognition of graphical primitives from raster images of documents, recognition of graphic symbols in charts and diagrams, CAD conversion of engineering drawings or diagrams, analysis of maps, charts, line drawings, tables or forms, 3-D model reconstruction from multiple 2-D views, etc.

In August 1995, IAPR's[1] technical committee on graphics recognition (TC10) organized the First International Workshop on Graphics Recognition, at the Penn State University's Scanticon Conference Center. The goal of the workshop was to bring together researchers from around the world to assess the state of the art in the above-mentioned topics. The workshop attendance was limited to 75 persons, to promote closer interaction among participants.

The workshop was organized into five sessions; each of them began with an invited talk assessing the state of the art, followed by short research presentations and concluding with a panel discussion, to identify important open research problems and suggestions for future research directions. These panel discussions proved to be very interesting, with much interaction between the authors and all the other participants.

A keynote talk on the commercial value of graphics recognition technology was presented by Dr. Clifford Kottman of Intergraph. He strongly emphasized the need for standards, this being one of the keys for graphics recognition methods developed by researchers to really meet industry's expectations.

This book is a collection of selected papers from the workshop. At the end of the workshop, we submitted all the papers presented at the workshop to the review of two workshop participants, who were asked to judge the contribution with respect to both the paper itself and the presentation and the subsequent discussions. As a result of this process, many of the papers were thoroughly revised and improved, so much so that we feel that this book has become much more than mere post-workshop proceedings. However, we have mostly limited our work to the scientific editing; basically, the authors provided camera-ready copies and we haven't done much copy-editing.

The book is organized into several main topics. The first deals with low-level processing, vectorization, and segmentation of scanned graphics documents. O'Gorman's paper gives an overview of the state of the art in this domain. The two following papers address the important topic of performance characterization and evaluation of low-level processing methods such as thinning or corner detection. *Vectorization*, i.e. raster-to-vector conversion, is an important step in graphics recognition. Although the basic methodology is well known, there is room for many improvements. Chhabra et al.'s

---

[1] International Association for Pattern Recognition.

paper describes a specific method for detecting horizontal lines in very large drawings, working directly on run length codes. Röösli and Monagan propose to add geometric contraint fitting to the vectorization process, in order to improve the quality of the result. Hori and Doermann propose a methodology for measuring the performances of vectorization algorithms. The next two papers deal with a specific application, that of analyzing forms. The last paper broadens the scope of low-level processing by proposing a method based on mathematical morphology for detecting dashed lines.

The second main topic of the book is symbol and diagram recognition. Blostein's introductory paper makes a broad review of the state of the art in this area. Messmer and Bunke present an efficient, error-tolerant method for finding subgraph isomorphisms between model graphs and a vectorized drawing; they also propose a learning algorithm for identifying new symbols which the system does not know yet. The two last papers present typical applications: recognition of logo-like graphics and interpretation of chemical structure diagrams.

A third area is that of map processing, i.e. the conversion of paper maps to some kind of GIS representation. Den Hartog et al. give an overview of work in this domain and present extensively their own research and propose a knowledge-base strategy which combines bottom-up and top-down processing. The next paper presents a method for region labeling in map processing. Myers et al. describe the approach they have developed for text and feature extraction. The last paper in this section addresses the crucial problem of the system design methodology in map processing, based on the authors' experience.

The interpretation of engineering drawings is also an important domain in graphics recognition. Four papers illustrate various aspects of this problem. Dori et al. describe the segmentation and recognition of dimensioning. Thomas et al. propose an approach combining low-level analysis and the use of higher-level information such as that conveyed by the textual content of the drawing, for instance. Capellades and Camps illustrate how functionality can be used in drawing interpretation, with a specific example, that of recognizing screws. Finally, Tomiyama and Nakaniwa propose a method for constructing 3D CAD models from three 2D views.

During the workshop, a contest to determine the best algorithm for detection of dashed lines in drawings was also organized. Formal performance evaluation protocols and metrics were used to test the accuracy of detection and representation of dashed lines. The paper by Kong et al. describes the performance analysis program developed for this contest. Many groups expressed interest in this contest and obtained test data but only one group completed their development and came to the workshop with a working program. The team headed by Dr. Dov Dori from Technion, Israel, received the award in this contest, and we have included in this book a paper written by this team and describing the program they submitted to the contest.

The book ends with a chapter summarizing the conclusions and recommendations from the numerous discussions and exchanges during the workshop. This chapter also includes more specific summaries of some of the sessions.

The workshop was supported by grants from the US Department of Defense and from industry. We owe special thanks to the contributing authors and to the reviewers of the papers. David Kosiba, who was responsible for local arrangements, worked hard to make everything run very smoothly and pleasantly. Many thanks also to Sarah Jane Barrier (Penn State University) and to Isabelle Herlich (INRIA Lorraine & CRIN/CNRS) for efficient help with all the secretarial duties, both for the organization of the workshop and during the preparation of this book.

The workshop was deemed very successful by the participants, and we plan to organize future workshops of the same kind; the next will most probably be held in Summer 1997.

March 1996   Rangachar Kasturi
                                                                   Karl Tombre

Workshop chairs

Rangachar Kasturi, U.S.A.

Karl Tombre, France

Program committee

Sergey Ablameyko, Belarus
Vincenzo Consorti, Italy
Dov Dori, Israel
Stephen Joseph, U.K.
Babu Mehtre, Singapore
Theo Pavlidis, U.S.A.

Atul Chhabra, U.S.A.
Luigi Cordella, Italy
Osamu Hori, Japan
Gerd Maderlechner, Germany
Lawrence O'Gorman, U.S.A.
Ken Tomiyama, Japan

Additional referees

Keiichi Abe, Japan
Adnan Amin, Australia
Abdel Belaïd, France
Dorothea Blostein, Canada
Maria Capellades, U.S.A.
Herbert Freeman, U.S.A.
Jurgen den Hartog, Netherlands
Ton ten Kate, Netherlands
Bin Kong, U.S.A.
Young-Bin Kwon, Korea
Michihiko Minoh, Japan
Gregory Myers, U.S.A.
Janet Poliakoff, U.K.
Markus Röösli, Switzerland
Shigeru Shimada, Japan

Gady Agam, Israel
Jason Balmuth, U.S.A.
Raymond Bennett, U.S.A.
Horst Bunke, Switzerland
Its'hak Dinstein, Israel
Edward Green, U.S.A.
Hidekata Hontani, Japan
In-Gyn Koh, Korea
David Kosiba, U.S.A.
Simone Marinai, Italy
Gladys Monagan, Switzerland
Thomas Nartker, U.S.A.
Arathi Prasad, U.S.A.
Eric Saund, U.S.A.
Arnold Smeulders, Netherlands

Table of Contents

Map processing

Engineering drawings

Dashed-line detection contest

Basic Techniques and Symbol-Level Recognition — An Overview

Lawrence O'Gorman

AT&T Bell Laboratories
Murray Hill, NJ, 07940-0636
email: log@research.att.com

Abstract. This is an overview paper that describes methods used in graphics recognition from the stage of the initial scanned image to that of the graphics features. The objective is to give context and background to the papers in this section — and, because it is introductory material — to the rest of the papers in this book.

Methods are described under the categories of: pixel-level processing, line-level processing, and feature detection. In the pixel-level processing section, thresholding, noise reduction, and compression are discussed. In the line-level processing section, thinning, chain coding, region detection, and polygonalization are discussed. And in the final section on feature detection, critical point detection, line and curve fitting, and shape recognition are discussed.

1 Introduction

The objective of graphics image analysis is to locate lines, symbols, regions, and other graphics components in images, for the ultimate objective of extracting the intended information — as a human would. A typical sequence of processing steps in graphics image analysis is the following: 1) data capture, 2) pixel-level processing, 3) line-level analysis, and 4) feature detection and recognition of text and graphics. This paper serves as an introduction to techniques for the first three items in this sequence.

Data capture is performed on a paper document usually by optical scanning. The resulting data is stored in a file of picture elements, called pixels. These pixels may have values: OFF (0) or ON (1) for binary images, 0-255 for gray scale images, and 3 channels of 0-255 color values for color images. At a typical sampling resolution of 300 dpi (dots per inch), an 8.5×11 inch page would yield an image of 2550×3300 pixels. For an engineering line drawing of size 34×44 inch and requiring high resolution of 1000dpi, the image is a huge $34,000 \times 44,000$ pixels. It is upon this raw pixel data that subsequent image processing steps are performed.

The pixel-level processing stage includes binarization, noise reduction, signal enhancement, and segmentation. The objective in methods for binarization is to automatically choose a threshold that separates the foreground graphics from the background of the page. Though graphics images are inherently binary in information, it is often advantageous to capture a gray-scale version of the image, then perform binarization using an adaptive algorithm (rather than to just use a fixed binarization level set in a scanner).

There may be noise in the captured image due to image transmission, photocopying, or degradation due to aging. A particular problem application for engineering drawings is blueprints, where the noise is often extreme. Specific filters for binary noise reduction — in particular salt-and-pepper noise — are appropriate for graphics images, and morphological filters are also designed for reducing this and other types of noise.

Signal enhancement is similar to noise reduction, but uses domain knowledge to reconstitute expected parts of the signal that have been lost. Signal enhancement is often applied to graphics components of document images to fill gaps in lines that are otherwise intended to be continuous.

Segmentation occurs on two levels for a graphics image. On the first level, text and graphics are usually discriminated for processing by separate methods: OCR for text and graphics analysis for graphics. On the second level, segmentation is performed on graphics by separating symbol and line components. For instance, in a page containing a flow chart with an accompanying caption, text and graphics are first separated. Then the text is separated into that of the caption and that of the chart. The graphics is separated into rectangles, circles, connecting lines, filled regions, etc.

The line-level analysis stage follows pixel-level processing. Line features describe each corner, curve, and straight line, as well as the rectangles, circles, and other geometric shapes. There are many methods for critical point detection (corners and curves) and for line and curve fitting. These are used for feature analysis of the line components. Shape description methods are used for filled regions.

The final stage of graphics analysis is the feature detection and recognition stage. Though this is not covered in this paper, the stage is described here for completeness. Components are assigned a semantic label and the entire document is described as a whole. It is at this stage that domain knowledge is used most extensively. The result is a description of a document as a human would give it. For an electrical circuit diagram, for instance, we refer not to lines joining circles and triangles and other shapes, but to connections between AND-gates, transistors and other electronic components. The components and their connections describe a particular circuit that has a purpose in the known domain. It is this semantic description that is most efficiently stored and most effectively used for common tasks such as indexing and modifying particular document components.

As we noted in the abstract, the objective of this overview paper is to give context and background to the papers in this section of the book. This is a representative sampling of common methods in a common sequence of processing for graphics recognition. This is comprehensive neither in scope nor depth; however, representative references accompany the descriptions of each method. Further depth and scope can be found in the tutorial textbook [1].

In the following sections, we will overview techniques in the following categories:

- Pixel-Level Processing — thresholding, noise reduction, compression
- Line-Level Processing — thinning, chain coding, region detection, and polygonalization
- Feature-Detection — critical point detection, line and curve fitting, and shape recognition.

2 Pixel-Level Processing

2.1 Thresholding

A gray-scale image has a range of intensities that is relatively large — often 256 levels — but may have far fewer levels of information. This page, for example, has two levels of information: the black text and the white background. However a gray-scale image of this page will have many more than two intensity levels due to non-uniform printing of characters and shadows caused by lighting effects. One of the most basic of image processing methods is binarization. This is a process of obtaining image regions corresponding to the two true levels of information from the many more intensity levels. This is also called bi-level thresholding because the process entails finding an intensity value that is said to be the threshold, where the range of intensities below corresponds to one information level, and the range above corresponds to the other information level.

If the pixel values of the components and those of the background are fairly consistent in their respective values over the entire image, then a single threshold value can be found for the image. This use of a single threshold for all image pixels is called global thresholding. For many documents, however, a single global threshold value cannot be used even for a single image due to non-uniformities within foreground and background regions. For example, for a document containing white background areas as well as highlighted areas of a different background color, the best thresholds will change by area. For this type of image, different threshold values are required for different local areas; this is adaptive thresholding.

For graphics recognition applications, both thresholding approaches are used. Global thresholding is preferable if the image conditions allow it, because the thresholding decision is based on larger statistics — the whole image. However, often for larger images such as large engineering drawings, the printing and lighting may not be uniform, therefore adaptive thresholding is more appropriate.

A number of global thresholding techniques determine classes by formal pattern recognition techniques that optimize some measure of separation. One approach is Otsu's method [2,3]. Calculations are first made of the ratio of between-class variance to within-class variance for each potential threshold value. The classes here are the foreground and background pixels and the purpose is to find the threshold that maximizes the variance of intensities between the two classes, and minimizes them within each class. This ratio is calculated for all potential threshold levels and the level at which the ratio is maximum is the chosen threshold. Another thresholding approach is by moment preservation [4]. For this method, a threshold is chosen that best preserves moment statistics in the resulting binary image as compared with the initial gray-scale image. These moments are calculated from the intensity histogram — the first four moments are required for binarization.

Another approach is the connectivity-preserving thresholding method [5]. This is global in that one threshold is found for the entire image, but it also combines a measure of local information, namely connectivity, upon which the threshold is determined. The method was designed in particular for document images, and tests show that the OCR rates derived from the thresholded images by this method were higher than those by other popular global thresholding methods.

A common way to perform adaptive thresholding is by analyzing gray-level intensities within local windows across the image to determine local thresholds [6,7,8]. White and Rohrer [9] describe an adaptive thresholding algorithm for separating characters from background. The threshold is continuously changed through the image by estimating the background level as a two-dimensional running-average of local pixel values taken for all pixels in the image. The main problem with any adaptive binarization technique is the choice of window size. The chosen window size should be large enough to guarantee that a large enough number of background pixels are included to obtain a good estimate of average value, but not so large as to average over nonuniform background intensities. However, often the features in the image vary in size such that there are problems with fixed window size. To remedy this, domain dependent information can be used to check that the results of binarization give the expected features (a large blob of an ON-valued region is not expected in a page of smaller symbols, for instance). If the result is unexpected, then the window size can be modified and binarization applied again.

A recent comparison of global methods for document images is in [5]. A recent comparison of methods for non-uniform document applications is in [10].

2.2 Noise Reduction

The first step after scanning is the reduction of noise in the image. For graphics images in which the information is binary, salt-and-pepper noise is the most prevalent. This noise appears as isolated pixels or pixel regions of ON noise in OFF backgrounds or OFF noise (holes) within characters and other foreground ON regions. The process of removing these is called "filling". It is important for a number of the techniques following preprocessing that the noise be reduced as much as possible. For storage of the images, noise reduction reduces the storage size. More importantly, for the recognition processes of OCR, symbol recognition, and line detection, the application of noise reduction can mean a substantial improvement in recognition results.

One common technique to reduce this noise is by mathematical morphology [11]. This is an iterative process. First, a chosen number of iterations of erosion, or reduction in size of image regions, is performed to eliminate small noise regions. Then the same number of iterations of dilation, or expansion in size of image regions, is performed to restore the size of the regions remaining after the small noise regions have been eliminated. There is a drawback in applying morphology in this way that sharp corners become rounded. A filter designed to maintain the original sharpness of the document components is the kFill filter [12]. This is a salt-and-pepper noise removal filter that errs on the side of maintaining text and graphics features versus reducing noise when those two conflict. The filter has a k parameter (the "k" in kFill) that enables adjustment for different text and symbol sizes, and image resolutions. This enables retention of small features such as periods and the stick ends of symbols.

2.3 Compression

The simplest, widely used lossless compression scheme for binary images is the fax standard, CCITT Group 3. This is a method combining 1-dimensional run-length coding of the rows along with Huffman coding of the run-lengths. A more efficient scheme is CCITT Group 4, which uses 2-dimensional run-length coding and Huffman coding. The current state-of-the art is the JBIG standard. This uses 2-D run-length coding and arithmetic coding and also enables progressive decoding. Rough comparative figures for the compression rates of these methods on a set of CCITT test images are the following. Group 4 compression can yield up to about half the size as for Group 3. JBIG can yield about a 25% improvement upon Group 4. (This is very dependent upon the types of images.)

Compression has not typically been described in the context of recognition techniques. However, the compression field has reached a point that very little further compression can be gained by the digital signal processing approaches currently used. The compression community is now looking at recognition techniques to make further compression gains. For instance, as a replacement for the current fax standard (CCITT Group 3), researchers are looking at recognizing similar symbol bitmaps on a page. This is a process akin to OCR, but without the final recognition decision. See [13,14,15] for references.

3 Line-Level Processing

3.1 Thinning

Thinning is an image processing operation in which binary valued image regions are reduced to lines that approximate the center skeletons of the regions. This is often done in the field of graphics recognition where many of the objects are lines. The purpose is to reduce thicker lines to chains of single pixels that can be easily traversed. This, in turn, facilitates further steps in analysis, such as line and curve detection and shape recognition.

A common thinning approach is performed iteratively; on each iteration every image pixel is inspected, and single-pixel wide boundaries that are not required to maintain connectivity or end lines are erased (set to 0 or OFF). This method consists of examining windows of 3×3 or 3×4 pixels throughout an image, and erasing the center pixel if the thinning criteria are met. The thinning criteria basically ensure that connectivity is maintained of lines and curves (i.e. lines are not broken or joined), and that line endings are not shortened.

Other methods have been proposed to thin with a fixed number of steps — not dependent on the maximum line thickness. For these non-iterative methods, skeletal points are estimated from distance measurements with respect to opposite boundary points of the regions. Some of these methods require joining the line segments after thinning to restore connectivity, and also require a parameter estimating maximum thickness of the original image lines to limit the search for pairs of opposite boundary points. All these algorithms can be performed in sequential, raster-scan order.

A recent survey paper on thinning method is [16]. Some references on particular methods are [17,18,19].

3.2 Chain Coding

A common method following thinning is chain coding. The thinned lines are stored as sequential lists of pixel locations. The chain representation is a more efficient storage mode than the pixel representation. A more important advantage in this context is that, since the chain coding contains information on connectedness within its code, this can facilitate further processing such as smoothing of continuous curves, and analysis such as feature detection of straight lines.

The most common chain coding method is to store, not pixel locations, but directions from one pixel to the next neighbor in the chain. A neighbor is any of the adjacent pixels in the 3×3 pixel neighbourhood around that center pixel. This is called the Freeman chain code [20]. A drawback of this simple chain code method is that topology — branching of lines — is not directly obtainable from the code. There are chain codes that preserve topology features, for instance [21,22]. These enable interconnections among lines to be determined easily.

3.3 Polygonalization

After chain coding, a common step is to perform polygonalization. This is a further data reduction step in which the chains are approximated by straight line segments. The objective is to approximate a given curve with connected straight lines such that the result is close to the original, but that the description is more succinct. The user can direct the degree of approximation by specifying some measure of maximum error from the original. In general, when the specified maximum error is smaller, a greater number of lines is required for the approximation. The effectiveness of one polygonalization method compared to another can be measured in the number of lines required to produce a comparable approximation, and also by the computation time required to obtain the result.

The iterative endpoint fit algorithm [23], is a popular polygonalization method whose error measure is the distance between the original curve and the polygonal approximation. The method begins by connecting a straight line segment between endpoints of the data. The perpendicular distances from the segment to each point on the curve are measured. If any distance is greater than a chosen threshold, the segment is replaced by two segments from the original segment endpoints to the curve point where distance to the segment is greatest. The processing repeats in this way for each portion of the curve between polygonal endpoints, and the iterations are stopped when all segments are within the error threshold of the curve.

The iterative endpoint fit algorithm is relatively time-consuming because error measures must be calculated for all curve points between new segment points on each iteration. Instead of this "overall" approach, segments can be fit starting at an endpoint, then fit sequentially for points along the curve [24,25,26,27]. That is, starting at an endpoint, a segment is drawn to its neighbouring point, then the next, etc., and the error is measured for each. When the error is above a chosen threshold, then the segment is fit from the first point to the point previous to the above-threshold error. This process is continued starting at this point as before. The error criterion used for this approach states that the straight line fit must be constrained to pass within a radius around each data point. Area between the curve and the polygonal fit can also be used as a measure of error [28] as well as the sum of square errors [29].

4 Feature Detection

4.1 Critical Point Detection

For many graphics drawings, all or most of the symbols can be drawn with straight line segments or circular curves. The corners between straight line segments and the boundary points of circular curves along lines are called critical points. Since corners and curves have intended locations, it is important that any analysis precisely locates these. It is the objective of critical point detection to do this precise recognition of locations — as opposed to polygonal approximation, where the objective is only a visually close approximation.

One approach for critical point detection begins with curvature estimation. A popular family of methods for curvature estimation is called the k-curvatures approach (also the difference of slopes approach) [30,31,32,33]. For these methods, curvature is measured as the angular difference between the slopes of two line segments fit to the data around each curve point. Curvature is measured for all points along a line, and plotted on a curvature plot. For straight portions of the curve, the curvature will be low. For corners, there will be a peak of high curvature that is proportional to the corner angle. For curves, there will be a curvature plateau whose height is proportional to the sharpness of the curve (that is, the curvature is inversely proportional to the radius of curvature), and the length of the plateau is proportional to the length of the curve. To locate these features, the curvature plot is thresholded to find all curvature points above a chosen threshold value; these points correspond to features. Then, the corner is parameterized by its location and the curve by its radius of curvature and bounds, the beginning and end transition points from straight lines around the curve into the curve. The user must choose one method parameter, the length of the line segments to be fit along the data to determine the curvature. There is a tradeoff in the choice of this length. It should be as long as possible to smooth out effects due to noise, but not so long so as to also average out features. That is, the length should be chosen as the minimum arclength between critical points.

Another method for critical point detection follows a different approach [34,35]. Instead of regulating the amount of data smoothing by choice of the line segment length fit to the data as above, the difference of curvatures is first measured on only a few points along the curve (typically three to five); then these local curvature measures are plotted on a curvature plot and a Gaussian-shaped filter is applied to smooth noise and retain features. Critical points are found from this curvature plot by thresholding as before. The tradeoff between noise reduction and feature resolution is made by the user's choice of the width of the Gaussian smoothing filter. Similarly to the length of the fit for the k-curvatures method above, the Gaussian filter width is chosen as wide as possible to smooth out noise, but not so wide as to merge adjacent features. The user must empirically determine a width that yields a good tradeoff between noise reduction and feature maintenance.

4.2   Line and Curve Fitting

Lines and curves, or portions of lines and curves, can be represented by mathematical parameters of spatial approximations, or fits, made to them. These parameters can be used in turn to identify features in the objects for subsequent recognition or matching. This representation by parameters instead of by the original pixels is useful because the parameters can provide a useful description of the objects as, for example: an object of four straight lines joined at 90° corners (a rectangle), or one with a specific type of curve that matches a curve of a particular object of interest. To aid the fitting process, we can use information from analyses already described to guide the fit, such as the polygonal approximations or locations of critical points. It is also helpful to have some *a priori* idea of the type of fit and the limited shapes of objects in an application to reduce computation and ensure better fits.

A simple way to fit a straight line to a curve is just to specify the endpoints of the curve as the straight line endpoints. However, this may result in some portions of the curve having large error with respect to the fit. A popular way to achieve the lowest average error for all points on the curve is to perform a least-squares fit of a line to the points on the curve. There can be a problem in blindly applying the least-squares method just described to document analysis applications. That is, the method is not independent of orientation, and as the true slope of the line approaches vertical, this method of minimizing y-error becomes inappropriate. A more general approach is to minimize error not with respect to x or y, but with respect to the perpendicular distance to the line in any orientation. This can be done by a line-fitting method called principal axis determination or eigenvector line fitting. This is further detailed in [36,37].

Especially for machine-drawn documents such as engineering drawings, circular curve features are prevalent. These features are described by the radius of the curve, the center location of the curve, and the two transition locations where the curve either ends or smoothly makes the transition to one or two straight lines around it [33,38]. One sequence of procedures in circular curve detection begins by finding the transition points of the curve, that is the locations along the line where a straight line makes a transition into the curve, and then becomes a straight line again. Next, a decision is made on whether the feature between the straight lines is actually a corner or a curve. Finally, the center of the curve can be determined with the information that, since each transition point is defined to be at a tangent to the curve, two perpendicular lines through these transition points will intersect at the curve center.

A different approach for line and curve fitting is by the Hough transform [39]. This approach is useful when the objective is to find lines or curves that fit groups of individual points on the image plane. The method involves a transformation from the image coordinate plane to parameter space. For a line, there are two parameters, slope and intercept, therefore the Hough space has dimensionality of 2. All points on the image plane are transformed to all possible points of slope and intercept values (or usually (ρ, θ), the polar coordinates) on the Hough plane. Peaks then indicate lines in the image plane with those parameter values. The method can be extended to circles and higher order curves, however this is at a much higher memory and computational expense for the higher dimensionalities.

4.3 Shape Recognition

In graphics applications, shape descriptors are most often used for recognition of non-elongated region objects, such as company logos and other symbols. (This is as opposed to line and curve features that are used for elongated objects as has been described above.) In this section, a sampling of shape descriptors are described. There are many shape descriptors, of which only a few can be mentioned in this section. A particular shape measure is chosen very much dependent upon the type of object it is to describe and on the types of objects among which it is meant to differentiate.

One of the simplest ways to describe shapes is by shape metrics. For instance, the area measurements (number of ON-valued pixels) of connected components of a document can be used to separate large symbols from smaller ones, or text from larger line graphics. Instead of counting up all the pixels in a connected component, a faster determination is the length of the contour, especially when the shape has already been stored as the chain code of its contour. This length measurement can be used to separate small regions from large ones. Of course, contour length and region area are not directly related — they also depend on other aspects of the shape. If the fatness versus thinness of a shape is an important discriminator, then compactness can be measured; that is the ratio of the square of the contour length over the area. This is small for a circular contour, and large for a long shape or a shape with many concavities and convexities. A common graphics entity that is separated using these size and shape metrics is lines, such as lines on forms. These are easily recognized and separated from other graphics and characters on the basis of being very long and narrow.

Another way to describe a shape is by its moments. These can be used to indicate the non-roundness, eccentricity, or elongation of the shape; for instance, if the second moments in x and y are similar in value, then the object is more round than otherwise. The third central moment gives a measure of the lack of symmetry of this eccentricity. Besides these central moments, moments of different orders can also be combined into functions that are invariant to other geometric transformations such as rotation and scaling; these are called moment invariants [40,41]. Moments are less useful to describe more complicated shapes with many concavities and holes; in these cases, specific features such as number of holes or thinned topology are more applicable. A company logo is a typical graphics that is often recognized based on distinctive moment or by topological features described below.

Shapes can be described by topological features, that is the number of holes and branches. For instance, the letter, "B", has two holes; and the letter, "P", has one hole and one branch. If a shape has many convexities, its skeleton will have many branches. For example, the thinned results of the symbols, "*" and "X" will have six and four branches respectively. The number of branches, the number of loops, the number of end lines, and the directions, lengths, etc. of each are all descriptive of the original shape.

The contour of a shape can be used for shape description. A global description of the contour can be determined by calculating the Fourier transform of the contour along its arc length. This gives a frequency domain description of the boundary in terms of Fourier coefficients, called Fourier descriptors. Lower order terms contain information of the large scale shape, that is its low-frequency curvature characteristics; and higher order terms contain information on sharper features, that is high frequency curvature characteristics. For further description of Fourier descriptors, see, for instance, [41]. A complicated diagram of a machine part might be adequately described for identification from other machine parts simply from its outline.

For more comprehensive reviews of shape description techniques, see [42,43]. For an extensive collection of papers on shape analysis, see [44].

5 Summary

We have presented an overview of techniques used in graphics recognition from the initial stage to the stage of scanned features.

References

1. L. O'Gorman, R. Kasturi, "Document Image Analysis", IEEE Computer Society Press, Los Vaqueros, CA, 1994.
2. N. Otsu, "A threshold selection method from gray-level histograms", IEEE Trans. Systems, Man, and Cybernetics, Vol. SMC-9, No. 1, Jan. 1979, pp. 62- 66.
3. S.S. Reddi, S.F. Rudin, H.R. Keshavan, "An optimal multiple threshold scheme for image segmentation", IEEE Trans. Systems, Man, and Cybernetics, Vol. SMC-14, No. 4, July/Aug, 1984, pp. 661-665.
4. W-H. Tsai, "Moment-preserving thresholding: A new approach", Computer Vision, Graphics, and Image Processing, Vol. 29, 1985, pp. 377-393.
5. L. O'Gorman, "Binarization and multi-thresholding of document images using connectivity", CVGIP: Graphical Models and Image Processing, Vol. 56, No. 6, Nov. pp. 494-506, 1994.
6. K.Y. Wong, "Multi-function auto-thresholding algorithm", IBM Technical Disclosure Bulletin, Vol. 21, No. 7, 1978, pp. 3001-3003.
7. R.G. Casey, K.Y. Wong, "Document analysis systems and techniques", in *Image Analysis Applications*, R. Kasturi and M.M. Trivedi (eds), Marcel Dekker, 1990, pp. 1-36.
8. M. Kamel, A. Zhao, "Extraction of binary character/graphics images from grayscale document images", CVGIP: Graphical Models and Image Processing, Vol. 55, No. 3, 1993, pp. 203-217.
9. J.M. White, G.D. Rohrer, "Image thresholding for optical character recognition and other applications requiring character image extraction", IBM J. Res. Development, Vol. 27, no. 4, July 1983, pp. 400- 411.
10. O. D. Trier, T. Taxt, "Evaluation of binarization methods for document images", IEEE Trans. PAMI, Vol. 17, No. 3, Mar. 1995, pp. 312-320.
11. R.M. Haralick, S.R. Sternberg, X. Zhuang, "Image analysis using mathematical morphology", IEEE Trans. PAMI, Vol 9, July 1987, pp. 532-550.
12. L. O'Gorman, "Image and document processing techniques for the RightPages Electronic Library System", Int. Conf. Pattern Recognition (ICPR), The Netherlands, Sept. 1992, pp. 260-263.

13. R. Aravind, G. L. Cash, D. L. Duttweiler, H-M. Hang, B. G. Haskell, A. Puri, "Image and video coding standards", AT&T Technical Journal, Jan-Feb., 1993, pp. 67-88.

14. JBIG, "Progressive bi-level image compression", ISO/IEC International Standard 11544, 1993.

15. I. H. Witten, A. Moffat, T. C. Bell, "Managing Gigabytes: Compressing and Indexing Documents and Images", Van Nostrand Reinhold Publ., New York, 1994.

16. L. Lam, S-W. Lee, C.Y. Suen, "Thinning methodologies — A comprehensive survey", IEEE Trans. Pattern Analysis and Machine Intelligence, Vol. 14, No. 9, Sept. 1992, pp. 869-885.

17. C.J. Hilditch, "Linear skeletons from square cupboards", Machine Intelligence 4, 1969, pp. 403-420.

18. N.J. Naccache, R. Shinghal, "SPTA: A proposed algorithm for thinning binary patterns", IEEE Trans. Systems, Man, and Cybernetics, Vol. SMC-14, No. 3, 1984, pp. 409-418.

19. L. O'Gorman, "k x k Thinning", CVGIP, Vol. 51, pp. 195-215, 1990.

20. H. Freeman, "Computer processing of line drawing images", Computing Surveys, Vol. 6, No. 1, 1974, pp. 57-98.

21. J.F. Harris, J. Kittler, B. Llewellyn, G. Preston, "A modular system for interpreting binary pixel representations of line-structured data", in Pattern Recognition: Theory and Applications, J. Kittler, K.S. Fu, L.F. Pau (eds.), D. Reidel Publishing Co. pp. 311-351, 1982.

22. L. O'Gorman, "Primitives Chain Code", in "Progress in Computer Vision and Image Processing", edited by A. Rosenfeld and L. G. Shapiro, Academic Press, San Diego, 1992 pp. 167-183.

23. U. E. Ramer, "An iterative procedure for the polygonal approximation of plane curves", Computer Graphics and Image Processing, 1:244-256, 1972

24. I. Tomek, "Two algorithms for piecewise-linear continuous fit of functions of one variable", IEEE Trans. Computer, Vol. C-23, No. 4, pp. 445-448, 1974

25. C. M. Williams, "An efficient algorithm for the piecewise linear approximation of planar curves," Computer Graphics and Image Processing, 8:286-293, 1978

26. J. Sklansky, V. Gonzalez, "Fast polygonal approximation of digitized curves," Pattern Recognition, 12:327-331, 1980

27. C. M. Williams, "Bounded straight-line approximation of digitized planar curves and lines," Computer Graphics and Image Processing, 16:370-381, 1981

28. K. Wall, P.E. Danielsson, "A fast sequential method for polygonal approximation of digitized curves," Computer Graphics and Image Processing, 28:220-227, 1984

29. Y. Kurozumi, W. A. Davis, "Polygonal approximation by minimax method," Computer Graphics and Image Processing, Vol. 19, 1982, pp. 248-264.

30. A. Rosenfeld, E. Johnston, "Angle detection on digital curves", IEEE Trans. Computers, 22: 875-878, Sept. 1973.

31. A. Rosenfeld, J.S. Weszka, "An improved method of angle detection on digital curves", IEEE Trans. Computer, Vol. C-24, 1975, pp. 940-941.

32. H. Freeman, L. Davis, "A corner-finding algorithm for chain-coded curves", IEEE Trans. Computer, Vol. C-26, 1977, pp. 297-303.

33. L. O'Gorman, "Curvilinear feature detection from curvature estimation", 9th Int. Conference on Pattern Recognition, Rome, Italy Nov., 1988, pp. 1116-1119.

34. H. Asada, M. Brady, "The curvature primal sketch", IEEE Trans. Pattern Analysis and Machine Intelligence, Vol. PAMI-8, No. 1, Jan., 1986, pp. 2-14.

35. A.P. Pridmore, J. Porrill, J.E.W. Mayhew, "Segmentation and description of binocularly viewed contours", Image and Vision Computing, Vol. 5, No. 2, May 1987, pp. 132-138.

36. R. O. Duda, P. E. Hart, *Pattern Classification and Scene Analysis*, Wiley-Interscience, New York, 1973, pp. 332-335.

37. D. H. Ballard, C. M. Brown, *Computer Vision*, Prentice-Hall, New Jersey, 1982, pp. 485-489.

38. P.L. Rosin, G.A.W. West, "Segmentation of edges into lines and arcs", Image and Vision Computing, Vol. 7, No. 2, May 1989, pp. 109-114.

39. J. Illingworth, J. Kittler, "A survey of the Hough transform", Computer Graphics and Image Processing, 44:87-116, 1988.

40. M. K. Hu, "Visual pattern recognition by moment invariants", in *Computer Methods in Image Analysis*, J. K. Aggarwal, R. O. Duda, A. Rosenfeld (eds.), IEEE Computer Society, Los Angeles, 1977.

41. A. K. Jain, *Fundamentals of Digital Image Processing*, Prentice Hall, New Jersey, 1989, pp. 377-381.

42. T. Pavlidis, "A review of algorithms for shape analysis", Computer Graphics and image processing", vol. 7, 1978, pp. 243-258.

43. S. Marshall, "Review of shape coding techniques", Image and Vision Computing, Vol. 7, No. 4, Nov. 1989, pp. 281-294.

44. C. Arcelli, L. P. Cordella, G. Sanniti di Baja (eds.), *Visual Form: Analysis and Recognition*, (Proceedings of the International Workshop on Visual Form, May, 1991, Capri, Italy), Plenum Press, New York, 1992.

An Alternative Approach to the Performance Evaluation of Thinning Algorithms for Document Processing Applications

L.P. Cordella and A. Marcelli

Dipartimento di Informatica e Sistemistica
Universita' di Napoli "Federico II"
Via Claudio, 21 - I 80125 Napoli (Italy)
cordel@nadis.dis.unina.it, marcelli@nadis.dis.unina.it

It is proposed that the performance of thinning algorithms be evaluated with reference to a task which is especially relevant in connection with the use of these algorithms in the application domain of document processing: decomposition of digital lines into meaningful parts. The stability of the decompositions obtained according to simple rules, within given classes of lines, is assumed as a performance index. Experimental results, obtained using the ETL1 database of handprinted characters, are presented, to demonstrate the representativeness of the considered parameter.

Keywords: Thinning, Performance evaluation, Document processing

1. Introduction

The most typical class of graphical objects making up documents is the one of thin lines. However, after digitization, printed and handwritten pages, line-drawings, geographical and technical maps, appear as sets of ribbon-like figures, that is, roughly speaking, of shapes whose length is appreciably greater than width. The human vision system and brain have developed enough skill to be able to ignore ribbon thickness, unless it becomes necessary in order to make a decision. Since in case of documents like the quoted ones this is quite uncommon, the first and most natural idea, when setting about processing this type of images by computer, was to perform a preliminary thinning of digital ribbons to unit width digital lines (skeletons).

Because of their potential applications, developing thinning algorithms has been a very popular practice during the last 25 years (see e.g. Lam[1]). The majority of the algorithms proposed in the literature belong to one of two groups. The first and wider group includes algorithms according to which the pixels of a binary digital figure are iteratively removed taking into account suitable deletion rules until no more pixels can be deleted without violating the rules. These essentially take care that figure topology is preserved, end points are not deleted and the skeleton is centered within the figure. According to the algorithms of the second group, the skeleton is made of the local maxima in a distance transform and of a few more pixels necessary to preserve connectedness. The presence of the local maxima guarantees the reversibility of the transformation which is often referred as Medial Axis Transformation. Distance functions of different types have been used to implement it in the discrete plane.

Independent of the technique adopted to obtain it, there are a number of desirable properties that the skeleton of a figure should exhibit depending on the considered class of figures, the application at hand and the way the skeleton has to be used. For description and recognition purposes, the main requirements are:
1. Preservation of figure topology.
2. Invariance with figure rotation. This property, important for thick figures, does not seem as much important for ribbon-like ones.
3. Preservation of shape properties of the figure. Shape fidelity is considered an especially important property for OCR and document analysis, to guarantee the reliability of the raster to vector conversion. Moreover, according to certain description and recognition approaches, it is required to have faithful representations of line images in terms of simple pieces and of their interrelationships.
4. Noise insensitivity.
5. Small computational cost. This is a critical point for several practical applications.

It is our opinion that the main requirement in case of ribbon-like figures remains shape fidelity. In fact, practically all the existing algorithms introduce shape distortions mainly at crossings and joints of strokes. It has to be noted that the most important of such distortions do not depend on the specific algorithm nor on the metric used to compute the skeleton, but on the fact that almost all the proposed algorithms try to implement directly or indirectly the Blum definition of skeleton in the real plane[2] or some variant of it. According to this model, the skeleton of a figure is the locus of the points that are centers of the maximal disks inscribed in the figure and are then equidistant from at least two separate boundary points. Following the definition, whenever a ribbon self-intersects or two ribbons join or cross, maximal disks of larger diameter than those inscribed in each single ribbon piece may be generated, so giving place to displacements of the local axis of symmetry and then to different types of spurious inflections of the skeleton. Examples of some common shape distortions are shown in figure 1.

The most natural approach to the problem of eliminating such shape distortions would be to avoid their generation during the process of thinning. This would imply either to change the reference model or to partly modify it: some attempts in this direction have been made in recent years[3-5]. An alternative approach is that of trying to correct the distortions after they have arisen, using the information preserved by the skeleton [6-8]. In Boccignone[8] it is shown that this is possible starting from Medial Axis Transforms: since they preserve the whole significant information about the original figure, it is relatively easy to locate skeleton parts possibly distorted and to access information useful to decide with good reliability if the distortion actually exists and how it should be corrected.

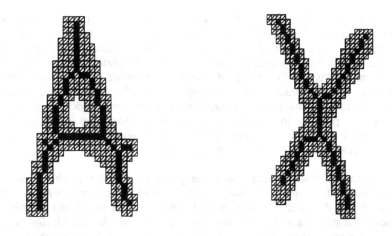

Fig. 1. Examples of typical shape distortions introduced by thinning algorithms.

Whatever the way of achieving a skeleton, the problem of assessing the suitability of the obtained transform holds. Although some papers qualitatively comparing thinning algorithms performance have been published since the '70s[9,10], in recent years the problem has been approached in a more quantitative way[11,12]. In Jaisimha[11] an error criterion function based on the Hausdorf distance is introduced, to measure the deviation between the output of a given thinning algorithm and an ideal skeleton based on the Blum's ribbon model. In Plamondon[12] it is suggested that the model with respect to which skeletons have to be compared, be generated by averaging those provided by a group of people, considered domain experts.

We propose an alternative approach to the quantitative performance evaluation of thinning algorithms, based on the above maintained considerations that, for document processing applications, shape fidelity is the most important feature of thinning and shape distortions possibly introduced by thinning mainly occur at crossings and junctions of strokes. In the following Section the rationale behind the approach will be briefly illustrated and experimental results supporting it will be presented.

2. The Performance Evaluation Method

In order to illustrate the method, let us consider the case of handprinted character recognition. According to many structural approaches to OCR, characters are preliminarily thinned to a unitary width digital line and then, generally after polygonal approximation of that line, decomposed into pieces; finally a description of the obtained parts and of their relations is forwarded to the classifier. Decomposition can be achieved in different ways, but, in the off-line case, a first split at branch points according to good continuity rules, seems one of the simplest solutions. However, the possibility of splitting joining lines into meaningful pieces according to good continuity rules is heavily affected by the distortions introduced by thinning. Although it has to be expected that, because of the large shape variability of hand-made characters, members of a same class may be decomposed in some different ways, thinning distortions considerably increase this variability. Therefore, in principle, it seems reasonable to judge the performance of a thinning algorithm on the basis of the amount of distortions introduced and, equivalently, on the basis of the correctness of the decompositions achievable starting from the obtained thin line. To make this approach operative and independent of the human judgment, we propose to use, as a performance index, the stability of the decompositions obtained, according to predefined splitting rules, for a set of characters that should be attributed to the same class.

Thinning and subsequent line decomposition are steps in the process of character recognition. One of the aims of every pattern recognition system is that of trying to absorb the variability within each class, without destroying information necessary to discriminate between members of different classes. Ideally, all the members of a same class should be decomposed, and therefore described, in the same way. Of course, there may be differently shaped samples within a same class, as in the case of characters, but decompositions of similarly shaped ones should be stable.

Let us consider a set of characters whose identity is known and the unit width digital lines obtained by applying a thinning algorithm to the character bit map. The proposed performance evaluation procedure can be outlined in the following way:
i) approximate the skeleton with a polygonal line. The tolerance of the approximation should be very small in order not to blindly destroy significant information. This step is convenient in order to apply good continuity rules in a simpler way;
ii) at every branch point of the skeleton, split the merging lines according to a set of good continuity rules. These can be of the type proposed in Boccignone[13] or as simple as: whenever three or more segments merge, keep together the most aligned pairs and split the other;
iii) for every class, compute the number of the obtained decompositions and the relative frequency of each of them. These parameters give a quite representative measure of the shape fidelity of the used thinning, as it has been experimentally confirmed.

The method has been tested on the ETL1 data base, including 1445 handprinted characters for every class. Several thinning algorithms were considered. To highlight the relevance of shape fidelity at joints and crossings of lines for evaluating the performance of a thinning algorithm for pattern recognition purposes, the results obtained with two algorithms are reported here. The first one[14] is representative of the large majority of algorithms which comply with the Blum's model of skeleton, while the second[8] attempts to correct shape distortions introduced during the thinning process. In the following they will be referred as Algorithm 1 and Algorithm 2 respectively. The polygonal approximation of the skeletons provided by the two algorithms for the same characters of figure 1 are shown in figure 2.

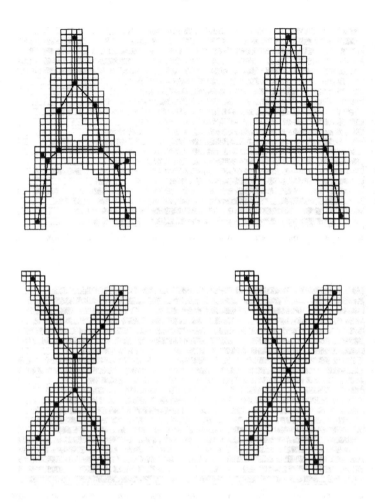

Fig. 2. Polygonal approximations of the skeletons provided by Algorithm 1 (left) and Algorithm 2 (right)

In figure 3, a bar diagram shows number N and percentage of occurrence P of the different character decompositions obtained with Algorithm 1 (top) and Algorithm 2 (bottom). On each horizontal bar, N is given by the number of sections, while length of these lasts represents P. The most right light gray section of some bars collects all the decompositions having P<2. Same texture of corresponding bars in the top and bottom diagrams corresponds to same decomposition. Because of space reasons, characters drawn with a simple line have been missed, so as W whose case is well represented by M. It is evident that the larger is N and the smaller is P for each decomposition, the less convenient is to use that algorithm in the framework of a structural pattern recognition process. It has to be taken account that, as already

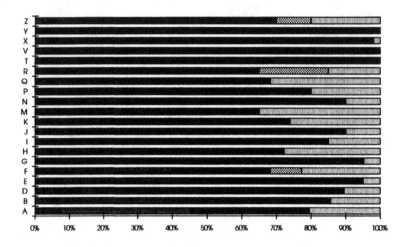

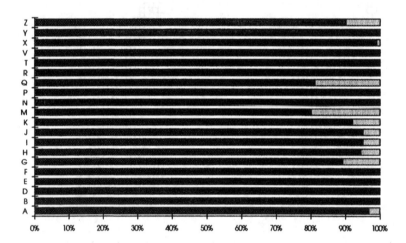

Fig. 3. Number and percentage of occurrence of the decompositions obtained starting from the skeletons given by Algorithms 1 (top) and 2 (bottom).

noted, within a same class there may be samples shaped in substantially different ways, thus necessarily leading to different decompositions. This is mainly the reason why, for each character, a few different decompositions with relatively high P value are obtained even with Algorithm 2. Therefore, the method seems more suitable for comparing algorithms performance than for absolute assessments, unless the test set is carefully prepared. In figure 4 the values of P for a subset of characters are shown, together with the corresponding actual decompositions, to give an idea of how different decompositions look like.

Finally, figure 5 exemplifies the confusion between classes that becomes

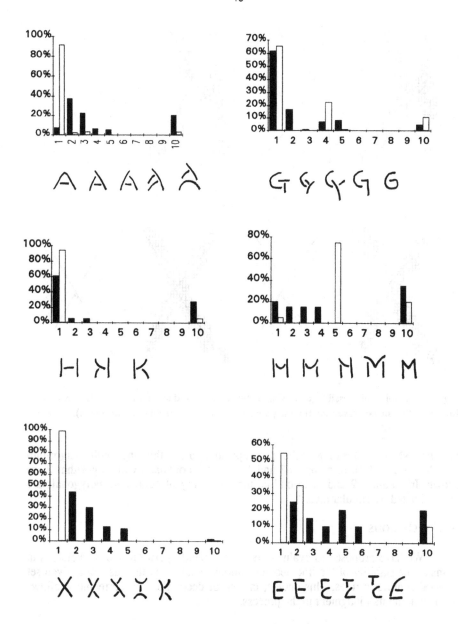

Figure 4: Percentage of occurrence P of the different decompositions for six character classes. Black bars refer to Algorithm 1, white bars to Algorithm 2. Bars denoted with numer 10 collect all the decompositions with P < 2. The set of the most frequent decompositions is sketched below each diagram.

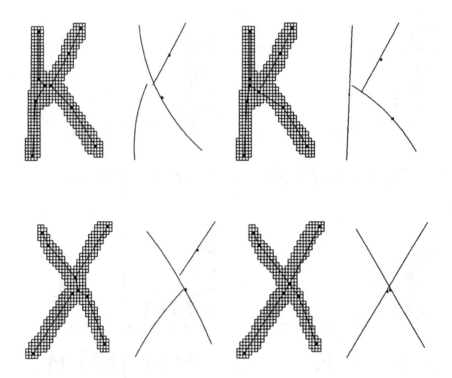

Fig. 5. Examples of possible confusion between classes due to thinning (left columns). Results with a shape preserving thinning are shown for comparison (right columns).

possible when performing the decomposition after thinning with Algorithm 1 (columns 1 and 2). In columns 3 and 4 the results obtained with Algorithm 2 are shown. In columns 2 and 4, to outline the possibility of confusion, polygonals are approximated by circular arcs.

3. Conclusions

We have proposed to evaluate the shape fidelity performance of skeletons in terms of the stability of the line decompositions obtained on the basis of a given set of good continuity rules. This choice, instead of decomposition correctness, allows us to avoid human judgment in the process.

The performance evaluation method proposed here differs from others recently proposed in the literature both for the features assumed to assess skeleton quality and because it doesn't require to use a term of comparison. Moreover, it can be applied to sets of lines intersecting each other.

The proposed performance evaluation method has been tested with different thinning algorithms on a large set of characters. The aim of the paper is that of illustrating the evaluation method and its implementation, rather than that of actually comparing specific thinning algorithms. However, the obtained results show that, according to the proposed performance index, algorithms based only on the Blum's model do not exhibit adequate shape fidelity for the purpose of document processing applications.

4. References

[1] L. Lam, S.W. Lee and C.Y. Suen, "Thinning methodologies - A comprehensive survey", *IEEE Trans. on Patt. Anal and Mach. Intell.*, vol. PAMI-14, no.9, 1992, pp.869-887.

[2] H. Blum, "A transformation for extracting new descriptors of shape", in: *Models for the Perception of Speech and Visual Form*, (W. Watjen-Dunn, ed.), MIT Press, Cambridge:MA, 1967, pp.362-380.

[3] V.K.Govindan and A.P. Shivaprasad, "A pattern adaptive thinning algorithm", *Pattern Recognition*, vol.20, no.6, 1987, pp.623-637.

[4] R.M. Brown, T.H. Fay and C.L. Walker, "Handprinted recognition system", *Pattern Recognition*, vol.21, no.2, 1988, pp.91-118.

[5] X. Li and A. Basu, "Variable-resolution character thinning", *Pattern Recognition Letters*, vol.12, 1991, pp.241-248.

[6] A. Sirjani and G.R. Cross, "On representation of a shape's skeleton", *Pattern Recognition Letters*, vol.12, 1991, pp.149-154.

[7] S.W. Lu and H. Xu, "False stroke detection and elimination for character recognition", *Pattern Recognition Letters*, vol.13, 1992, pp.745-755.

[8] G. Boccignone, A. Chianese, L.P. Cordella and A. Marcelli, "Using skeletons for OCR", in: *Progress in Image Analysis and Processing*, (V.Cantoni et al eds.), World Scientific Publishing Co., Singapore, 1990, pp.275-282.

[9] H. Tamura, "A Comparison of Line Thinning Algorithms from a Digital Geometry Viewpoint", *Proc. 4th ICPR*, Kyoto (JAPAN), 1978, pp.39-52.

[10] C:J:Hilditch, "Comparison of Thinning Algorithms on a Parallel Processor", *IVC*, vol.1, no.3, 1983, pp.115-132.

[11] M.Y. Jaisimha, R.M.Haralick and D. Dori, "A Methodology for the Characterization of the Performance of Thinning Algorithms", *Proc. ICDAR '93* Tsukuba (JAPAN), October 20-22, 1993, pp. 282-286.

[12] R: Plamondon, M: Bordeau, C: Chouinard, and C.Y. Suen,"Validation of Preprocessing Algorithms: A Methodology and its Application to the Design of a Thinning Algorithm for Handwritten Characters", *Proc. ICDAR '93*, Tsukuba (JAPAN), October 20-22, 1993, pp. 287-290.

[13] G. Boccignone, A. Chianese, L.P. Cordella and A. Marcelli, "Recovering Dynamic Information from Static Handwriting", *Pattern Recognition*, vol. 26, no.3, 1993, pp.409-418.

[14] C: Arcelli and G: Sanniti di Baja, "A Thinning Algorithm based on Prominence Detection", *Pattern Recognition*, vol.13, no.3, 1981, pp.225-235.

Comparison of Methods for Detecting Corner Points from Digital Curves

Tokuhisa Kadonaga and Keiichi Abe

Dept. of Computer Science, Shizuoka University, Hamamatsu 432, Japan

Abstract. Though many methods have been proposed for the detection of dominant points from digital curves, comparisons of general performance have seldom attempted, nor the advantages and disadvantages of each method have been investigated. As a case study of performance evaluation of image processing algorithms, this report describes the results of comparing 11 dominant point detection methods, from two aspects: (1) invariance of the set of detected points, and (2) evaluation by human subjects. Only the corner points detected are evaluated and compared.

1 Introduction

Feature point detection from a given linear curve plays a crucial role in decomposing or describing the curve. We consider here only digital curves of one-pixel width, i.e., curves obtained by connecting the grid points of a square grid. Among the feature points to be detected, end points, branching points, and crossing points are detected by checking their connectivity numbers. However, feature points with connectivity number two, that is, corner points, inflection points and transition points should be detected otherwise. Here we define the transition point as a point where the curve transit smoothly from a straight line segment to a curved segment or vice versa.

There have been proposed dozens of method to extract corner points from digital curves; some of them claim to be able to extract inflection points and transition points. However, most of these methods were only proposed, without vast comparison of performance against other methods [13,14]. In consequence, little is clear about which methods work comparatively well in general, or what merits or demerits the individual method has.

We report here the result of comparison of methods proposed for extracting corner points from digital curves. The viewpoint of comparison is twofold: (a)invariance of extracted points under rotation, size change, and reflection; (b)subjective evaluation and comparison with the results made by human beings.

One big confusion exists in the distinction between feature point detection and polygonal approximation of digital curves. The feature points should be detected from the viewpoint of describing input figures well by linear segments and circular arcs via those points. Intuitive importance might also be emphasized in this approach. On the other hand polygonal approximation makes a stress on how to decrease approximation errors. To an extreme the extracted points should not be necessarily critical points in intuitive sense. Thus the authors think that approximation errors used in [7] is not an appropriate criterion for feature point detection.

The methods compared are the followings:

(1) A method using n-codes [1,13]
(2) A method using slope information [2]
(3) Freeman-Davis method [3]
(4) Rosenfeld-Weszka method [4]
(5) Koyama et al. method [5]
(6) Arcelli et al. method [6]
(7) Teh-Chin method [7]
(8) DOS (Difference of Slope) method [8]
(9) Fischler-Bolles method [9]
(10) Beus-Tiu method [10]
(11) Held et al. method [11]

Though some of these methods claim that they can detect inflection points as well as corner points, we compare here only the performance in extracting corner points, that is, the points where the digital curvature takes a large absolute value.

2   Invariance under Rotation, Size Change, and Reflection

The set of extracted corner points should not be changed under rotations, size changes (enlargement or shrinking), and reflections of input curves. Since rotations and size changes cause a small deviation in digital curves in general, the invariance should be more or less an approximate one. In case of size changes one may think the set of dominant points to be detected should be different, because the figures becomes larger or smaller, that is, its details are described finer or coarser. We assumed, however, that the set should be identical under the range from 1/2 to 2 of size changes we investigated here. The analysis was made by using four curves shown in Figure 1.

Comparison was made by investigating what percentage of extracted points from the transformed figures is the same as the points extracted from the original figures. The invariance was measured by using

$$E = (1 - \frac{\#(\bar{A} \cap B) + \#(A \cap \bar{B})}{\#(A) + \#(B)}) \times 100(\%)$$

where,

A: the set of extracted points in the original figure,
B: the set of extracted points in the transformed figure,
#(.): the number of the points in a set.

If the two sets of extracted points are completely identical, then the figure of merit E becomes 100%, and if they are totally different, it is 0%.

Most of the 11 methods used in the experiments have one or more parameters to be adjusted. They were set as follows: for each method one experimenter extracted the corner points in the original four figures with varying the parameter(s), and decided the best value(s) to be used. This set of values was also applied to the transformed figures.

Now we will remark some notes for each of three figure transformations.

2.1 Invariance under Rotation

The curves in Figure 1 were rotated by 15, 30, 45 degrees and the invariance of extracted points were looked into. We tried the following three methods for rotating input figures.

(1) Applying of rotation matrix and resampling
(2) Rotation by shearing
(3) Rotation of continuous (analog) curve on the scanner

As the method (3) was found least affected by the sampling noise, it was adopted. Thus the rotation angles are approximate ones.

2.2 Invariance under Size Change

The curves were enlarged or shrunk by $2.0, 1.5, 1.2, 3/4, 2/3, 1/2$ times. Actually, the original figures were sampled much finer than shown in Figure 1, and they were re-sampled in different pitches corresponding to the magnification factors.

2.3 Invariance under Reflection

Reflection of the input figure is equivalent to inverting the direction of curve tracing, say, clockwise to counterclockwise. In this experiment ideally the two sets of extracted points should be identical. In practice some methods extracts corner points shifting by one or a few points, when the curve is reflected or the tracing direction is inverted. Often this is because in the intermediate stage of extraction a group of consecutive points is detected as a corner and then some tie-breaking rule is applied to select one from the group. As it seems unavoidable, we modified Eq.(1) by permitting a shift of detected corner point up to one point in both direction. The modified measure is denoted as E'.

2.4 Results of Experiments on Invariance

The values of E or E' for each method averaged over four input figures are shown in Figure 2.

As for rotational invariance, good results were obtained for n-code and DOS methods, poor for Teh method. When the size of the figures was changed, n-code and Arcelli methods yielded relatively invariant results, while the results of Freeman and Beus methods changed a lot. In reflecting the input figures, perfect unchanged results were obtained with Rosenfeld, Koyama, Fischler methods; some points shifted to their direct neighbor in n-code method.

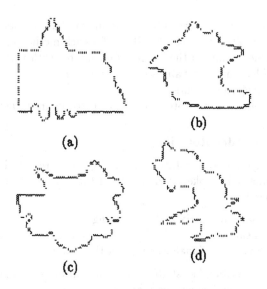

Fig. 1. Curves used in invariance tests

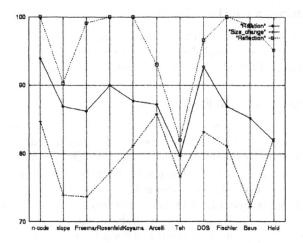

Fig. 2. Evaluation of invariance

3 Evaluation by Human Subjects

It is reasonable that even a method is quite good in invariance tests described in the last section, it should be called poor if it extracts too many corner points or misses too many points to be extracted referring to human intuition. Therefore in the second part of our experiments, we compared the 11 methods from the viewpoint of how exactly they extract the set of points which humans want to be extracted. The parameter(s) of each method, if exists, were adjusted so that it produces subjectively best extraction (observed by one experimenter), depending on each input curve.

Two criteria for evaluation were considered:

(a) Subjective evaluation by human subjects
(b) Comparison with the set of points which human subjects extracted

3.1 Subjective Evaluation by Human Subjects

Ten digital curves shown in Figure 3 was used and subjective evaluation of the results of the aforementioned 11 methods on those curves was made by 18 subjects. The grades for evaluation are:

5: excellent, 4: good 3: so-so, 2: a little poor, 1: poor

As in Arcelli method and Held method the dominant points are detected hierarchically, the individual subject was asked to choose the best level and put a grade for that level of result.

The average grades over 18 subjects are depicted in Figure 4. The figure is divide into two to show 11 lines corresponding to 11 methods. Its horizontal axis stands for the input curve number, with the rightmost being the average over 10 curves. The grade of each method varies from input to input, but in overall Rosenfeld's method was the best and got 4.11 in averaging over 10 curves. The second was n-code method with average score 4.03, while the worst was Teh's with 2.09. The rest formed a cluster with scores ranging from 3.49 to 3.14.

3.2 Comparison with the Extracted Points by Human Subjects

Extraction of dominant points by human subjects. The same 18 subjects was asked to extract corner points from the same 10 digital curves and their results were compared with those of the machine methods. The request was to mark the points to be absolutely extracted, as well as the points which may or may not be extracted. The number of points to be extracted is not specified, that is, the subjects are requested to extract as many points as he or she feels reasonable. An example of human extraction is illustrated in Figure 5. Then the same results of 11 methods as the ones used in the subjective evaluation were compared with the extracted results of human subjects.

Comparison of the results of machine algorithms and human extraction. In order to compare the results of 11 machine algorithms with the results of extraction by human subjects, five ways of scoring the machine results were tried. In any of the scorings, the higher the score is, the better the method is. Because the highest and the lowest values vary from a scoring scheme to another, linear transformations were applied to normalize the highest value to 5 and the value 0 to 1. In some scoring the raw value can be negative and so the normalized value can be less than 1.

Scheme 1

In this scheme the score is decided as the sum of the scores for each point detected by a machine algorithm:

+**2**: A point which more than or equal to a half of the subjects extracted as mandatory. (In other words, we regard it as a dominant point to be detected mandatory.)

  0: Except the points above, a point which more than or equal to a half of the subjects extracted as mandatory or optional. (In other words, we regard it as a dominant point to be detected optionally.)

−1: Except for the above two cases, a point which at least one subject extracted as mandatory.

−2: Other points.

Scheme 2

The score is the sum of the following scores times the number of subjects who extracted that point as mandatory or optional.

+**1**: A point extracted as mandatory.

  0: A point extracted as optional.

−1: Others.

For example, consider a point detected in some algorithm. If 6 subjects extracted that point as mandatory, 5 as optional, and 7 did not extracted the point at all, then the score is $1 \times 6 + 0 \times 5 + (-1) \times 7 = -1$. The total score is the sum of such values for all points detected by the algorithm.

Scheme 3

This scoring scheme is to allow one point shift of detected points. We append the following to Scheme 1.

+**1**: A direct neighbor of a point which more than or equal to a half of the subjects extracted as mandatory.

Scheme 4

Scheme 1 may be too strict to the extra detected points. Human beings tend to feel it more serious to miss a point to be detected than to extract excessive points. In order to reduce the penalty relative to the gain, the score −2 in Scheme 1 is changed to −1.

Scheme 5

This scheme is obtained by applying the same penalty reduction mentioned above to Scheme 3.

Results of scoring. All of the results produced by the five scoring schemes showed qualitatively similar tendency to each other and also to the subjective evaluation. The results of scoring scheme 1 and 3 are shown in Figures 6 and 7.

3.3   Correlations between Subjective Evaluation and Scoring

As mentioned above, the results of subjective evaluation and the score compared to human extraction were qualitatively similar. To confirm the similarity quantitatively, we calculate the correlation coefficients between the grade derived in subjective evaluation and the scores calculated by five schemes and depict them in Table 2. Here r stands for the correlation coefficient calculated over 110 samples, that is, 11 methods times 10 input curves. r_{ave} denotes the correlation coefficient between the averaged grades or scores over 10 curves, i.e., between the values in the rightmost columns of Figures 4, 6 and 7.

For instance, The correlation between the subjective grades and the scores in Scheme 5 was 0.618 using the values per methods per curves, but 0.960 considered only averages over 10 curves. More or less the similar values can be seen for any of the scoring scheme. Thus the correlation is moderate if we deal with the inputs independently, but quite significant if considered as the whole.

The high correlations in the averaged case are important because it makes us possible to make a rough evaluation of the results obtained by a new method or by a method treated in this paper but with different values of parameters, without making subjective experiments. (Note that it is very hard to make a *fair* subjective comparison in different times.)

Table 1. Correlations between subjective evaluation and scores

Scheme	1	2	3	4	5
r	0.625	0.616	0.688	0.552	0.618
r_{ave}	0.952	0.947	0.950	0.945	0.960

4   Discussions and Conclusion

Eleven corner point detection methods are compared from the viewpoints: (1) invariance under rotation, size change, and reflection of the input curves, and (2) subjective evaluation and comparison with the dominant points extracted by human subjects.

In invariance tests, the method using n-code produced generally best invariant results under three kinds of figure transformations. >From the performance compared with the human subjects, Rosenfeld-Weszka method was the best. Slightly worse was n-code methods. As the latter is simple and fast, it is recommendable unless the quality of extraction is critically important. It may be sarcastic that the best two methods are rather classic ones. Of course some of the methods compared are intended to describe the input figure as a whole, rather than just to detect corner points; consequently the way of comparison may not

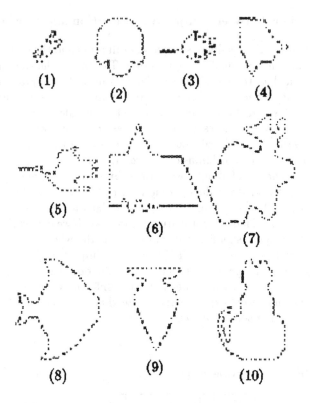

Fig. 3. Curves used in evaluation by human subjects

be fair for those methods. As for the subjective evaluation and comparison with the dominant points extracted by human subjects, we could confirm that they yielded qualitatively a similar tendency.

It should be emphasized that the evaluation in this paper is restrictive, as every performance evaluation of image processing methods is, the authors believe. The relative goodness or poorness of specific method is validated only under the condition we examined here. Over-generalization of the conclusion would be risky.

There are a lot of things to be done further.

1. The way of specifying dominant points to be extracted by human subjects should be improved. For instance, a complaint was heard about how to specify in case that either of two points should be extracted. The scoring scheme should be also modified to deal with such a case. This has been done already and the general tendency was not changed.

2. The robustness of the methods in question should be also studied. As mentioned before, the parameters of the methods were set optimally for each of the input. Another comparison should be made by fixing the parameters independently of the inputs, and also the performances of the methods should

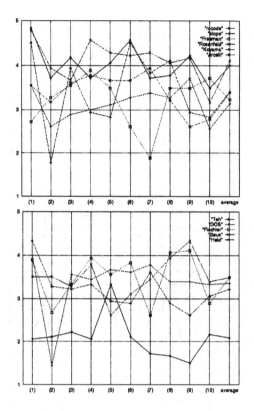

Fig. 4. Human subjective evaluation

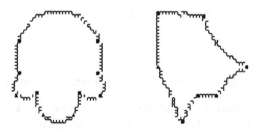

!|: Points to be extracted

"%: Points to be optionally extracted

Fig. 5. Example of human extraction

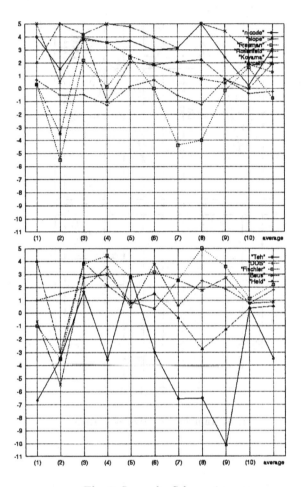

Fig. 6. Scores by Scheme 1

be investigated under the deviations of parameter values from the optimum.

3. Dominant point detection methods are increasing every year. Some promising methods have not yet been included in this comparison, because of shortage of time and manpower, or difficulty of programming them. Sending their programs to the authors would be welcome.

4. Maderlechner pointed out that from practical view comparisons with input figures of higher resolution is more important [15]. This is another thing to be done further. Line/arc fitting approach [16] seems promising in this case.

5. Computational complexity is another aspect for comparison, which was partly reported in our preliminary paper [13].

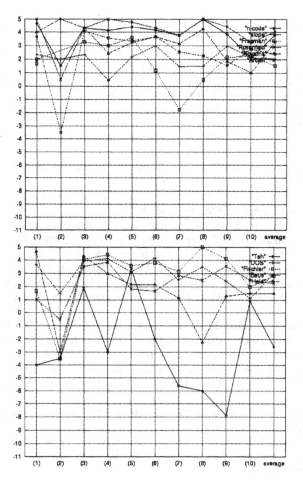

Fig. 7. Scores by Scheme 3

Acknowledgement

This work was partially supported by a Grant-in-Aid for Priority Areas from the Ministry of Education, Science, and Culture, Japan (No.06212209) and Real World Computing Partnership. The authors are also grateful to Dr. L. O'Gorman for his program and to the participants of GREC'95 and the referee for their valuable comments.

References

1. Gallus, G. and Neurath, P.W.: Improved computer chromosome analysis incorporating preprocessing and boundary analysis. Phys. Med. Biol. **15** (1970) 435-445
2. Ibaraki, T.: B.S. Thesis, Dept. of Computer Science. Shizuoka University (1991) (in Japanese)

3. Freeman, A. and Davis, L.S.: A corner-finding algorithm for chain-coded curves. IEEE Trans. on Computer **C-26** (1977) 297-303

4. Rosenfeld, A. and Weszka, J.S.: An improved method of angle detection on digital curves. IEEE Trans. on Computer **C-24** (1975) 940-941

5. Koyama, T., Shiono, M., Sanada, H. and Tezuka, Y.: Corner detection on thinned pattern. IEICE Technical Report **IE80-119** (1981) (in Japanese)

6. Arcelli, C., Held, A. and Abe, K.: A coarse to fine corner-finding method. Proc. IAPR Workshop on Machine Vision Applications (1990) 427-430

7. Teh, C.-H. and Chin, R.T.: On the detection of dominant points on digital curves. IEEE Trans. on PAMI **11** (1990) 859-872

8. O'Gorman, L.: Curvilinear feature detection from curvature estimation. Proc. 9th Int'l Conf. on Pattern Recognition (1988) 1116-1119

9. Fischler, M.A. and Bolles, R.C.: Perceptual organization and curve partitioning. IEEE Trans. on PAMI **PAMI-8** (1986) 100-105

10. Beus, H.L. and Tiu, S.S.H.: An improved corner detection algorithm based on chain-coded plane curves. Pattern Recognition **20** (1987) 291-296

11. Held, A., Abe, K. and Arcelli, C.: Towards a hierarchical contour description via dominant point detection. IEEE Trans. on SMC **24** (1994) 942-949

12. Kasturi, R. Siva, S. and O'Gorman, L.: Techniques for line drawing interpretation: an overview, Section 3.2. Proc. IAPR Workshop on Machine Vision Applications, (1990) 151-160

13. Abe, K., Morii, R., Nishida, K. and Kadonaga, T.: Comparison of methods for detecting corner points from digital curves - a preliminary report. Proc. 2nd Int'l Conf. on Document Analysis and Recognition (1993) 854-857

14. Legault, R. and Suen, C.Y.: A comparison of methods of extracting curvature features. Proc. 11th IAPR Int'l Conf. on Pattern Recognition **III** (1992) C-134 - C-138

15. Maderlechner, G.: Comment at the International Workshop on Graphics Recognition (1995)

16. Sugimoto, K. and Tomita, F.: Boundary segmentation by detection of corner, inflection and transition points. Proc. IEEE Workshop on Visualization and Machine Vision (1994) 13-17

Detection of Horizontal Lines in Noisy Run Length Encoded Images: The FAST Method

Atul K. Chhabra[1], Vishal Misra[1,2], and Juan Arias[1]

[1] NYNEX Science & Technology, 500 Westchester Avenue
White Plains, NY 10604 USA

[2] *Summer Intern at* NYNEX Science & Technology
Graduate Student at Department of Electrical and Computer Engineering
University of Massachusetts at Amherst, Amherst, MA 01003 USA

Abstract. We present a fast method for finding horizontal lines in run length encoded images. The method was motivated by the need for quick and reliable detection of horizontal lines in an interactive drawing conversion system for telephone company drawings. At the core of the algorithm are the processes of filtering run lengths, assembling filtered run lengths, generating top silhouette, and thresholding the gradient of the top silhouette to extract one horizontal line at a time. The method is robust in the presence of distortion; it can tolerate significant skew and warping, both local and global, and can bridge significant breaks in lines without too many false positive lines.

1 Introduction

Line detection is a low level process that is essential for any graphics interpretation system. In order to make the high level interpretation more accurate, line detection techniques have to be very reliable. When dealing with engineering drawings, the scanned images tend to be very large. Therefore, for building an interactive drawing interpretation system, very fast line detection techniques are needed.

This paper addresses the problem of finding horizontal lines in drawing images that have a high degree of distortion. The method operates on run length representation of images. The steps involved are filtering run lengths, assembling the filtered run lengths, generating the top silhouette of the assembled runs, and thresholding the gradient of the top silhouette to extract one line completely visible (unoccluded) from the top. The assemble, silhouette, and threshold steps are applied repeatedly until all horizontal lines in the image are detected. The Filter, Assemble, Silhouette and Threshold (FAST) method is very quick and is robust in the presence of noise [1].

The FAST method has been developed for use in an interactive drawing conversion system for telephone company drawings. Telephone companies use a huge number of engineering drawings in their central office operations. A vast majority of these drawings contain tabular or quasi-tabular data. Some examples

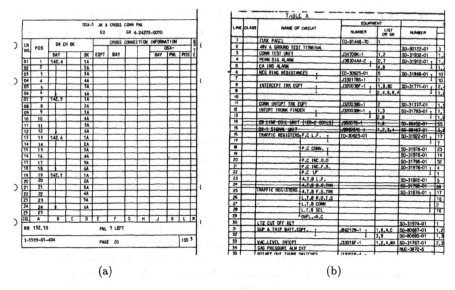

(a)   (b)

Fig. 1. Examples of telephone company drawings with structured information. The figure shows small sections of (a) an assignment drawing and (b) a wiring list drawing.

of such structured drawings are assignment drawings, wiring lists (see Fig. 1), front equipment drawings, and distributing frame drawings (see [1]). The data in these drawings has traditionally been stored on sheets of mylar because the process of storage and update dates back to the days when computers were not commonly used in telephone company operations. We are working on a system to semi-automatically convert the structured drawings into computer data. As can be seen in Fig. 1, the structured drawings, which account for a majority of the drawings, consist of lines that are mainly horizontal and vertical. In this work, we will present a special algorithm developed for detecting horizontal lines in the drawings.

Over years and decades of use, the mylar sheets go through copiers, drafting equipment, and filming equipment. The heavy use causes the mylars to develop extensive local warping and shearing effects and other distortion. Scanning introduces more distortions such as global skew, speckles, and faint scanning in certain regions of the drawings. Because of the large size of the mylar drawings, a small misalignment of the drawings during the scanning process causes the horizontal and vertical lines to appear very skewed. Because of the simultaneous presence of uneven warping, it is not possible to correct the skew to make the lines exactly horizontal and vertical. With age, patches of ink begin to fall off from the mylars because of wear and tear. This results in fragmented lines. Very often, text touches the lines. Due to these reasons, it is not possible to use an off the shelf vectorization package. We need a system that can tolerate severe warp and skew, text touching lines, and some fragmentation in the lines.

Traditional straight line detection algorithms, such as vectorization by thinning [2, 3], polygonal approximation [4, 5] and the Hough transform [6, 7, 8], do not work well on such distorted images. Moreover, these methods are computationally very expensive when applied to large images such as those for the E-size drawings (approximately 3' x 4'). The FAST horizontal line finding method, in contrast, can tolerate significant warping and skew, both local and global. This method operates on the run length description of a binary image. The method does not need the bitmap representation. This saves memory and computational cost. Pavlidis [9] proposed a vectorization method that works on compressed line adjacency graphs (c-LAGs) derived from run length images. The method requires building the c-LAGs, analyzing the paths, applying the width change criterion and collinearity of centers criterion, vectorization, and merging of vectors. The intermediate steps can be very expensive for noisy images and the vectorization may be orientation dependent. Dori, Liang, Dowell, and Chai [10, 11] developed a sparse pixel based line finding method called Orthogonal Zig Zag (OZZ). The OZZ method first finds horizontal and vertical lines using a modified OZZ-like move and then uses OZZ moves to find slanted lines. This is the fastest bitmap based method we have encountered. However, even when tailored to work well for only horizontal lines, the method takes much longer than the FAST method and makes lot of mistakes in noisy images. Besides, OZZ has too many parameters to adjust in order to work well on a given class of images.

Another application in which line finding is important is forms processing. However, the objective of our work is different from that of forms processing systems. The goal of forms processing is form identification and registration for data extraction. Casey, Ferguson, Mohiuddin, and Wallach [12] use horizontal and vertical lines on a form image for form identification and for registration. This requires matching the line descriptions to a set of form models. Line finding using run length encoded images is alluded to in [12]; the algorithm is not described and no timing or performance statistics are provided. Taylor, Fritzson, and Pastor [13] use a trainable neural network for form identification; they use corner features as inputs to the neural network classifier. Subsequent image registration gives enough information for extracting data entries on the form. Interpretation of telephone company tabular drawings presents a different set of problems altogether. There is a very large variety of tabular documents. Even within a document type, the individual tables have a large variability in their structure. The identification of table type is impractical. Therefore, the goal is to interpret the structure of a free format tabular or quasi-tabular drawing, and to extract, recognize, and relate data entries in very noisy and large document images. In order to find the structure, we need to discover all horizontal and vertical lines in the drawings. Due to the large size of the drawings, we use run length encoding to represent an image. Corner finding, the technique used in [13], can be applied easily to raw pixel images, not to run length encoded images.

In the rest of the paper, we present the FAST horizontal line detection algorithm in detail, illustrate the method on some scanned images, and compare results with other methods.

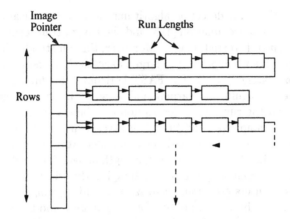

Fig. 2. Run length description of the image. The vertical array on the left is a 1-d array of pointers to the first run length in each scan line. Run lengths in a scan line are stored as a linked list. The linked lists of run lengths for consecutive scan lines are linked together.

2 The FAST Horizontal Line Finding Method

The algorithm proposed here can find horizontal lines in the presence of warping and in the presence of significant skew. The actual amount of skew that the algorithm can tolerate is given by[3] $\pm\sin^{-1}(MinLineThickness/FilterWidth)$.

The algorithm works directly on run length encoded images. The run length description of an image consists of a one-dimensional array of pointers to the data type run-length. The length of this array is equal to the image height. The elements in this array point to the first run length in the corresponding scan line of the image. Run lengths in a scan line are stored as a linked list. In addition, the last run length of each scanline is linked to the first run length of the next scanline. This provides a linked list representation of all the run lengths in an image while providing a quick access to the run lengths for any desired scan line. The description is depicted graphically in Fig. 2. The FAST algorithm consists of the following steps:

1. Filter: For run lengths lying on the same scanline, close any gaps smaller than *CloseGap* pixels, i.e., merge the run lengths less than *CloseGap* pixels apart. Next, filter out run lengths that are less than *FilterWidth* long. In doing so, we assume that all horizontal lines are longer than *FilterWidth*. We discuss the selection of *FilterWidth* in Sect. 3 below.

   During this step, run lengths with small breaks between them are connected together. Subsequently, small run lengths are discarded. Run lengths that remain after filtering are stored in the linked list structure of Fig. 2.

---

[3] All italicized words in this paper refer to parameters in the algorithm.

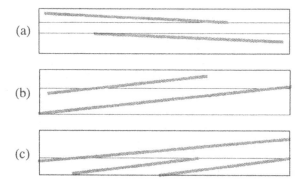

Fig. 3. Typical situations encountered in the grouping of lines in Step 2 of line detection. (a) When there is very limited skew and when the horizontal lines are well separated vertically, the run lengths comprising the two lines are clearly separated in the linked list representation. (b) With a larger skew and/or lesser vertical spacing between horizontal lines, the run lengths corresponding to distinct lines are intertwined in the linked list representation of the image, i.e., run lengths in the same scan line belong to more than one 'horizontal' line. (c) With large skew and/or little vertical spacing between horizontal lines, the run lengths corresponding to the three distinct lines are intertwined in the linked list representation of the image.

2. <u>Assemble:</u> Now the run lengths need to be assembled into horizontal lines. Starting from the first run length in the linked list, we group together all the run lengths in that scan line and in subsequent scan lines until we come across a vertical gap of two scan lines with no run lengths in those scan lines. This group of run lengths could consist of several horizontal lines. In the ideal case, when horizontal lines are perfectly horizontal and straight, this step could lead to a grouping together of horizontal lines that are collinear. In real images with skew and warping, this step could lead to grouping together of several long horizontal lines that lie above or below each other in addition to lines that may be collinear (see Fig. 3).

3. <u>Silhouette:</u> The next step is to find the top profile of the grouped run lengths. We traverse the linked list representation of the assembled run lengths and create the top profile or silhouette of those portions of all horizontal lines in the grouping that are visible from the top (i.e., not occluded by other horizontal lines above them). If there is no "on" pixel in a particular portion of the profile, the profile is set to -1 at that point.

   If all horizontal lines in the grouping have no horizontal overlap, then breaks in this top silhouette can be used to extract individual lines from the grouping. In practice, several horizontal lines in the grouping may have horizontal overlaps. Therefore, we need to figure out which part of this top silhouette corresponds to a horizontal line that is completely unoccluded

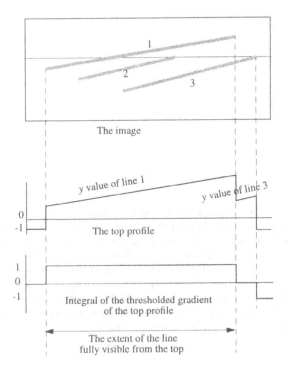

Fig. 4. The FAST algorithm to detect a horizontal line.

from the top. This is accomplished in the next step.

4. Threshold: We traverse the top profile computed in Step 3 from left to right and find the derivative of the profile. If the derivative is below a threshold of a few pixels, we treat the derivative to be zero. If the derivative is above the threshold, we only keep its sign and ignore the magnitude (i.e., we set the derivative to -1 or +1). We then compute the integral of this thresholded derivative function along the image width. Actually, the derivative and integral computation is done in a single step. In this integral, the extent of a contiguous region with the highest value corresponds to a horizontal line that is completely unoccluded from the top.

In Fig. 4, we see a portion of an image showing three separate lines with one line fully visible from the top, one partly occluded, and one fully occluded. Below that is the top profile for this section of the image. The following graph shows the integral of the (thresholded) gradient of the top profile, revealing the unoccluded line as the one with the highest integral.

5. Extract & Iterate: Using the information obtained from Step 4, we extract one real horizontal line from the grouping, i.e., we find all run lengths that

belong to this line. We determine which run lengths belong to this line by traversing all the run lengths in the grouping and checking which run lengths are within a distance *MaxLineThickness* from the top profile. Next, we compute the top profile as in Step 3 and the derivative and integral as in Step 4 for the remaining run lengths in the grouping. From this we obtain description of another horizontal line. We repeat this step until all run lengths in the grouping have been accounted for.

At this stage, we can optionally investigate the few lines in the grouping to check if any of the lines are collinear with small breaks (larger than *CloseGap*) and, if so, connect them.

6. <u>Iterate:</u> Next, we go back to Step 2 and iterate until all the run lengths that remain after Step 1 are accounted for. This gives us descriptions of all the horizontal lines present in the image irrespective of some warping and skew.

The procedure described above requires much less computation than line finding using Hough transform, vectorization by thinning, polygonal approximation, or the horizontal bar finding of the OZZ[4] method.

The FAST method, as described above, cannot detect differing line widths. However, it is possible to extend the FAST method to do so. While extracting one horizontal line at a time from a grouping of run lengths (Step 5), one could compute the bottom profile of the lines. The difference between the top and the bottom profiles could then be used as a measure of line width.

3 Effect of Parameters on Performance

The FAST algorithm has three adjustable parameters – *CloseGap, FilterWidth*, and *MaxLineThickness*. *CloseGap* and *FilterWidth* are used in the filtering step. For run lengths on the same scan line, any gap smaller than *CloseGap* is closed and subsequently any run length smaller than *FilterWidth* is discarded. *CloseGap* shouldn't be set too large, else gaps between run lengths belonging to text are closed and parts of text strings get interpreted as lines. *FilterWidth* should be set to at least the width of the largest text font in the image. If the inter-character spacing in any part of an image is very low (the extreme case being touching characters), it is desirable to set *FilterWidth* to about two to three times the width of the largest font. Else, parts of text strings get interpreted as lines On the other hand, if *FilterWidth* is set too high, small horizontal line segments in the image are not detected. This problem is more acute in images that have very thin lines and large skew (recall that the FAST method can tolerate skew of up to $\pm \sin^{-1}$ (*MinLineThickness/FilterWidth*)). If the skew is too large and the minimum line thickness in the image too low, then a physical line is composed of several short run lengths in different scan lines. If the size of the run lengths becomes comparable to *FilterWidth*, they are filtered out and the line is not detected.

---

[4] In the rest of the paper, OZZ refers to only the horizontal bar detection of the orthogonal zig zag algorithm [10, 11].

Any two run lengths that have a horizontal overlap but are vertically further apart than *MaxLineThickness* would not be considered part of the same line. For drawings having many different line thicknesses, we choose *MaxLineThickness* to be the maximum of the various thicknesses. If an image contains lines that are thicker than *MaxLineThickness*, then the FAST method detects multiple lines within each such thick line. On the other hand, if we keep the *MaxLineThickness* value too large, any two lines falling completely within a band *MaxLineThickness* wide would be merged together into a single line. Therefore, it is important to carefully choose the value of *MaxLineThickness*.

4 Experiments

We present results of applying the FAST method to several very noisy telephone company drawings and some medical insurance claim forms. All the images were scanned at 300 dpi. In our experiments, we used *CloseGap* of 5 pixels. *FilterWidth* was set to about two and a half times the maximum character width in the respective images, with one exception (see below), i.e., *FilterWidth* = 80 pixels for drawing images and 70 for form images. *MaxLineThickness* was set to 15 for drawings images and 4 for form images. We compare the results for some images with the OZZ method. Except for one image (as noted below), we set the *minWidth* parameter of the OZZ algorithm to two pixels.

Figure 5(a) shows part of a wiring list tabular drawing scanned at 300 dpi. This drawing has little or no skew. Local warping effects, although not present in the horizontal lines, are clearly visible in the vertical lines. Figure 5(b) shows the horizontal lines detected using the FAST method. The line detection and on-screen rendering took only 0.31 seconds on a 70 MHz Sun SparcStation-10 for the 2959×3858 pixel image.

In Fig. 6(a), we show part of an assignment table drawing scanned at 300 dpi. This drawing has a significant skew (approximately 2.5°). Figure 6(b) shows the horizontal lines detected using the FAST method. The line detection and on-screen rendering took 0.66 seconds on a 70 MHz Sun SparcStation-10 for the 2520×3394 pixel image.

The timings for the FAST method in Fig. 5 and 6 are faster than reported earlier in [1]. The speed up was obtained by more efficient coding.

Table 1 reports the timings of the FAST method for several other images. Also shown in Table 1 are the results for the OZZ method (only the horizontal bar detection) at two different levels of the *ScreenSkip* parameter. We compare our results with the OZZ method because it is the fastest line finding method reported in literature. The OZZ code we used was tailored to find only horizontal bars. As apparent from the table, the FAST method not only finds horizontal lines much quicker, it also requires much less memory. This is by virtue of the run length image representation used by the FAST method. The OZZ method used the raw image. Further, as discussed below, the quality of horizontal line detection of the FAST method is much better than that of the OZZ method.

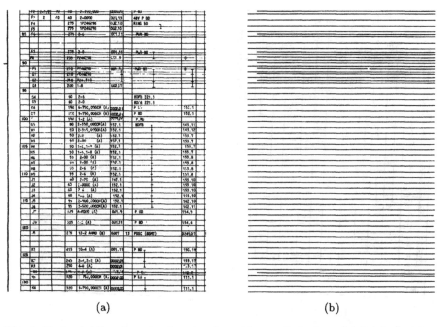

(a)   (b)

Fig. 5. Horizontal line detection in a drawing with very little skew: (a) Original wiring list image (width 2959 pixels, height 3858 pixels) and (b) the detected horizontal lines. For this image, after the filtering step, 796 run lengths remained for further processing. It took 0.31 seconds for horizontal line detection and rendering on a 70 MHz Sun Sparc-10.

The OZZ method has several deficiencies. OZZ is very sensitive to noise (see Figs. 7, 8, and 9). When horizontal lines are drawn through text, as in Fig. 7, OZZ tends to break up the lines due to the width change criterion of OZZ. OZZ works faster for larger values of the *ScreenSkip* parameter. However, with large *ScreenSkip*, it often misses significant parts of horizontal lines, particularly the right ends of horizontal lines (see Fig. 8). This happens when the line width at the last sampled point on a line is significantly different from the average line width over the rest of the line. The line gets broken here and the remaining line segment is never visited (because *ScreenSkip* is large). With smaller values of *ScreenSkip*, OZZ does not miss lines; however, smaller *ScreenSkip* introduces lots of undesirable noise lines from text.

As can be observed in Table 1, the computation time of the OZZ method doesn't scale linearly with the *ScreenSkip* parameter. For the large assignment drawing, the time required with *ScreenSkip* = 5 is almost the same as the time required with *ScreenSkip* = 10. This happens because, for this image, OZZ spends most of the time trying to merge fragmented horizontal bars. The number of horizontal bars found in the two instances is not very different.

For the dental claim form, OZZ takes much longer with a higher *ScreenSkip*,

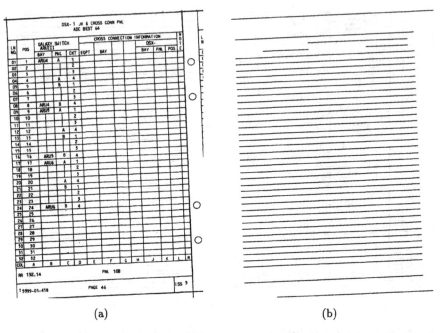

(a)   (b)

Fig. 6. Horizontal line detection in a drawing with significant skew: (a) Original skewed DSX assignment table image (width 2520 pixels, height 3394 pixels) and (b) the detected 'horizontal' lines. For this image, after the filtering step, 3658 run lengths remained for further processing. It took 0.66 seconds for horizontal line detection and rendering on a 70 MHz Sun Sparc-10.

which is counter intuitive. The reason for this is obvious upon looking at Fig. 9. Fig. 9(c), which corresponds to the larger *ScreenSkip*, has a much more noisy result, particularly in the shaded region of the image. More detected horizontal bars imply more time spent on attempts to merge the bars. Apparently, with the *ScreenSkip* of 10 pixels, a significant number of sampled pixels in the shaded region of the image turn out to be 'on'. This leads the OZZ algorithm into believing that there are horizontal lines in the shading. In contrast, the FAST algorithm will never falsely interpret a shaded region as having horizontal lines unless the gaps between the 'on' pixels of the shading are consistently less than the *CloseGap* parameter. The lines in this image are two to three pixels thick. The spacing between the double lines is two pixels. For the FAST method, we set *MaxLineThickness* for this image to four pixels. With *MaxLineThickness* set correctly, the FAST method is able to detect double line even with such a small spacing between the lines. With OZZ, it is hard to correctly detect double lines this close because they get merged during the merging step of OZZ.

The image of Fig. 7(a) contains very thin lines (one pixel thick in some places) and has a significant skew (1.5°). In order to get continuous lines, *FilterWidth* was set to a few pixels more than the maximum character width in the image

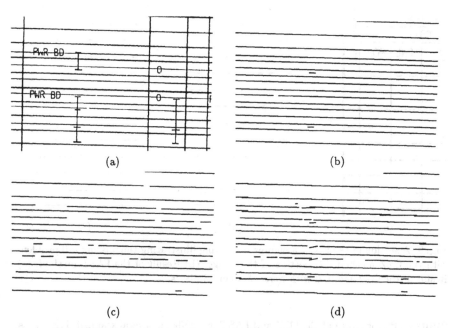

Fig. 7. The FAST method and OZZ as applied to a skewed wiring list drawing. The limits of the line finding methods are tested by several very thin and noisy lines. A section of the image is shown in (a). The result of the FAST method is shown in (b) and that of OZZ with *ScreenSkip* of 10 and 5 is shown in (c) and (d), respectively. Due to some very thin lines in the image, we set the *minWidth* parameter of OZZ to one pixel.

(see the discussion in the previous section). While applying the OZZ method to this image, the parameter *minWidth* was set to one pixel.

Although the OZZ method does not perform well for horizontal line finding in very noisy images, it does very well, in our experience, for general line finding in cleaner images (images having thick continuous lines of uniform width).

Janssen [14] surveyed the existing vectorization techniques. The timings reported in [14] for images of complexity comparable to the simplest images we use in our experiments are of the order of a minute (for finding all, lines not just horizontal).

GTXRaster CAD PLUS [15], a commercial vectorization software, running on a 120MHz Pentium PC with 32M RAM, and 120M swap space, took 2 minutes for vectorizing the image of Fig. 5(a) and 30 seconds for the image of Fig. 6(a). Of course, in this time, the software finds not only horizontal lines, but vectorizes the entire image (including text).

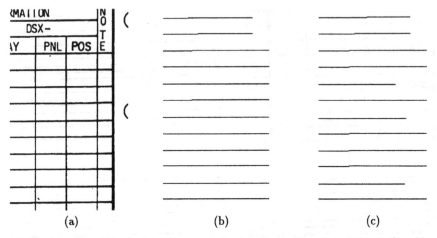

Fig. 8. A section of the E-size assignment drawing illustrating the missing tail problem with the OZZ method. (a) Section of the image. (b) Result of the FAST method. (c) Result of the OZZ method with *ScreenSkip* of 25 pixels.

Table 1. Performance of the OZZ and FAST methods on several scanned documents

		OZZ			FAST	
		Skip 5	*Skip* 10			
	Image Size	Time	Time	Mem.	Time	Mem.
Document Image	(pixels)	(sec)	(sec)	(MB)	(sec)	(MB)
Large Assignment Drawing	10144 × 6802	14.51	13.91	75.0	1.63	10.0
DSX Assignment Table	2360 × 3316	1.46	0.85	8.5	0.18	1.6
Skewed Assignment Table	2520 × 3394	2.97	1.37	9.0	0.66	1.7
Wiring List	2959 × 3858	2.44	1.97	12.5	0.31	1.8
Skewed Wiring List	2968 × 3858	4.27	4.73	13.0	0.55	2.0
Dental Claim Form	2784 × 3363	2.97	5.32	11.0	0.38	2.0
Insurance Claim Form	2784 × 3363	2.48	1.47	11.0	0.31	1.6

5 Summary

We presented and illustrated the FAST method, a quick and efficient method for detecting horizontal lines in noisy run length encoded images. We presented test results and comparisons with existing methods. By dealing with the run length description of an image, and by filtering out unnecessary run lengths, the FAST method drastically cuts down the storage requirement and the computational effort required for locating horizontal lines. On noisy document images, the quality of results and the speed of the FAST method are superior to the OZZ method, the fastest existing method for line finding.

6 Acknowledgements

The authors wish to thank María A. Capellades of the Penn. State University for providing us with her implementation of the OZZ algorithm.

References

1. Chhabra, A.K., Surya, S., and Misra, V.: Fast detection of horizontal lines in telephone company drawings. In *Proceedings of the IAPR International Workshop on Graphics Recognition*, pages 13–22, University Park, PA, August 1995.
2. Lam, L., Lee, S.W., and Suen, C.Y.: Thinning methodologies – A comprehensive survey. *IEEE Transactions on Pattern Analysis and Machine Intelligence*, 14(9):869–885, 1992.
3. Filipski, A.J. and Flandrena, R.: Automated conversion of engineering drawings to CAD form. *Proceedings of the IEEE*, 80(7):1195–1209, 1992.
4. Ramer, U.: An iterative procedure for the polygonal approximation of plane curves. *Computer Graphics and Image Processing*, 1:244–256, 1972.
5. Douglas, D.H. and Peucker, T.K.: Algorithms for reduction of the number of points required to present a digitized line or its caricature. *The Canadian Cartographer*, 10(2):112–122, 1973.
6. Hough, P.V.C.: Methods and means for recognizing complex patterns. U.S. Patent 3,069,654, 1962.
7. Duda, R.O. and Hart, P.E.: Use of the Hough transform to detect lines and curves in pictures. *Communications of the ACM*, 15(1):11–15, 1972.
8. Turolla, E., Belaid, Y., and Belaid, A.: Form item extraction based on line searching. In *Proceedings of the IAPR International Workshop on Graphics Recognition*, pages 262–271, University Park, PA, August 1995.
9. Pavlidis, T.: A vectorizer and feature extractor for document recognition. *Computer Vision, Graphics, and Image Processing*, 35:111–127, 1986.
10. Dori, D., Liang, Y., Dowell, J., and Chai, I.: Sparse-pixel recognition of primitives in engineering drawings. *Machine Vision and Applications*, 6:69–82, 1993.
11. Chai, I. and Dori, D.: Orthogonal zig-zag: An efficient method for extracting bars in engineering drawings. In Arcelli, C., Cordella, L.P., and Sanniti di Baja, G., editors, *Visual Form*, pages 127–136. Plenum, New York, 1992.
12. Casey, R., Ferguson, D., Mohiuddin, K., and Wallach, E.: Intelligent forms processing system. *Machine Vision and Applications*, 5(3):143–155, 1992.
13. Taylor, S.L., Fritzson, R., and Pastor, J.A.: Extraction of data from preprinted forms. *Machine Vision and Applications*, 5:211–222, 1992.
14. Janssen, R.D.T.: *The application of model based image processing to the interpretation of maps.* PhD thesis, Delft University of Technology, Delft, the Netherlands, 1995.
15. GTX Corporation, Phoenix, AZ. *GTXRaster CAD PLUS^{TM} Software, Version 2.6*, 1994.

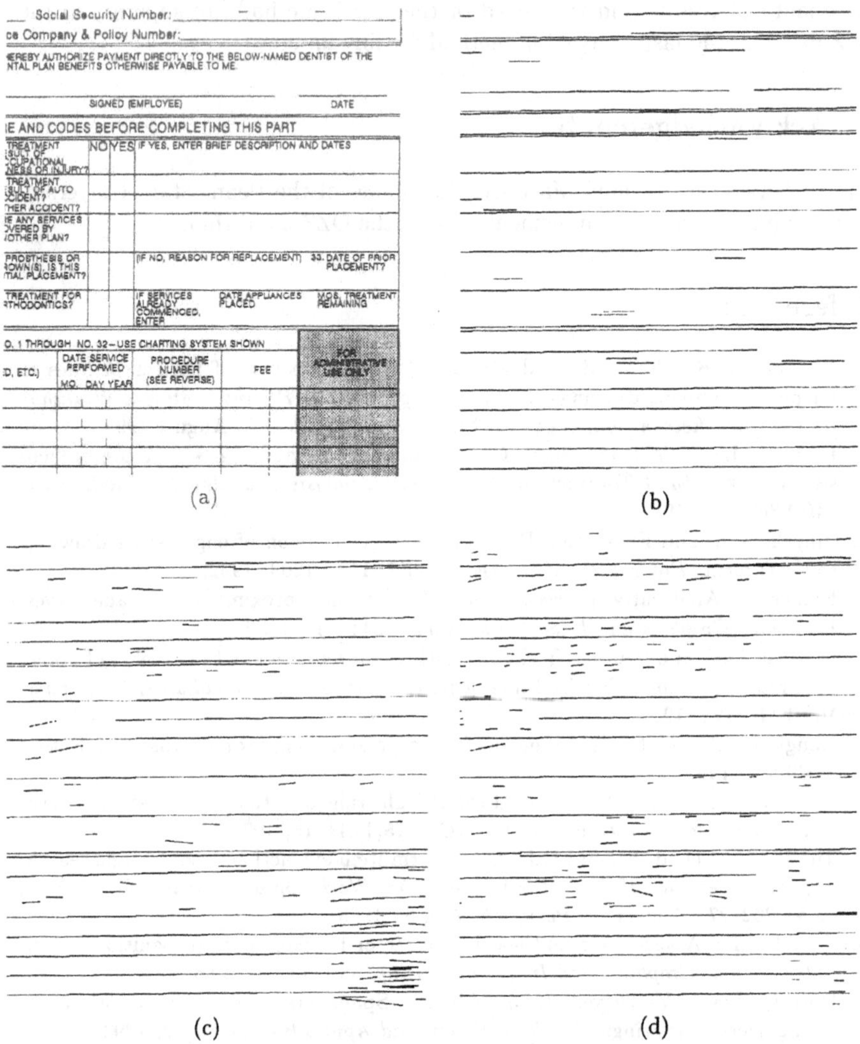

48

Fig. 9. The FAST method and OZZ as applied to a dental claim form. The limits of the line finding methods are tested by the several double lines and the shaded region. A section of the image is shown in (a). The result of the FAST method is shown in (b) and that of OZZ with *ScreenSkip* of 10 and 5 is shown in (c) and (d), respectively.

Adding Geometric Constraints
to the Vectorization of Line Drawings

Markus Röösli and Gladys Monagan
Institut für Informationssysteme
Swiss Federal Institute of Technology (ETH)
ETH-Zentrum, CH-8092 Zürich, Switzerland
roosli@inf.ethz.ch   gladys.monagan@inf.ethz.ch

Abstract. We consider the problem of recognizing line drawings like cadastral maps or technical drawings. We examine the vectorization (raster-to-vector process) of systems for solving this type of recognition problem and propose a solution that identifies and localizes graphical primitives. We claim that geometric constraints should be brought into the vectorization process and we give details on how we combine the vectorization and a constraint fitting for graphical primitives.

1 Introduction

The goal of a graphics recognition system for technical drawings and maps is to interpret line drawings so that the interpretation corresponds to the drawing meant by the draftsman. We are primarily interested in the interpretation of cadastral maps. We claim that such an interpretation system must deliver very accurate results. If the final interpretation of our maps is not precise enough, the user will refuse to use the system altogether.

The first step after scanning, and possibly some image processing, is a raster-to-vector conversion (e.g. [1] or [3]). If this raster-to-vector conversion is not accurate enough, irreparable errors will be introduced into the interpretation process: inaccuracies which cannot be corrected even at a later stage with additional knowledge.

The raster-to-vector conversion that we consider extracts graphic primitives like line segments and circular arcs from the raster image. What must be delivered is the number of primitives, the type of each primitive, its line width, its endpoints, and if an arc, its center point as well.

The land surveyors who are interested in our system have told us that if we cannot guarantee that two lines are parallel (as in the case of a wall drawn on a cadastral map) that they are not interested in our interpreted output. Thus, we require of an interpretation system that it support relations among primitives like:

- right angles between two primitives
- parallel line segments
- a common tangent between two primitives at the connection point

All these geometric relations can be expressed by defining constraints for tangents of the primitives determined by their endpoints. So, we claim that the vectorization has to support constraints for tangents.

In this paper, we describe our solution for a vectorization which includes a constraint fitting to deliver the required accurate primitives. We then conclude with a discussion about how to use this type of vectorization.

2 Vectorization

The purpose of the vectorization process is to deliver a set of primitives by translating information about the linear structures in the raster level up to the vector level. In model object recognition [2], one speaks of the recognition process consisting of *identifying* and *localizing* a model object in the scene. Similarly for graphics recognition, during the vectorization process, the number and types of primitives are *identified* and in addition, the primitives are *localized*.

In a similar way, Simon [6] looks for regular features as a first step, and then for singular features in a second step. In our case, regular features correspond to instances of line segments or circular arcs, and singular features to the endpoints of these segments and arcs.

Thus, we establish that a vectorization system can consist of two main steps. The first step is a segmentation which produces a kind of raw primitives (Fig. 1.) and the second step consists of fitting the raw primitives to generate the final primitives.

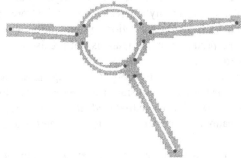

Fig. 1. Raw primitives. Note how the endpoints of the raw primitives are not connected.

2.1 First Step: Segmentation

During the segmentation, parts of the raster image are assigned to be raw primitives. The raw primitives do not have meaningful connections nor are the relations between the primitives reliable. For our implementation of such a segmentation, we use a data structure that assigns parts of the raster image to each raw primitive. We keep "pointers" from the raw primitives back to the raster image to be able to evaluate the recognition quality of a primitive. More details about the raster image representation and the segmentation process are given in [5].

2.2 Second Step: Error Fitting

In this step, the raw primitives are fitted to generate the final primitives. Usually a least squares fitting optimizes the raw primitives on their data sets (i.e. on the points representing the corresponding raster component) with the boundary conditions that the endpoints of adjacent primitives be connected.

2.3 Difficulties

If one implements this two step vectorization, difficulties arise while:

- determining the endpoints of the primitives accurately, especially when the part of the raster image which is being segmented corresponds to a continuous curve,
- localizing short primitives reliably, short in the sense that they correspond to small lines in the image or to short arcs,
- imposing geometric relations on primitives (i.e. constraints for the tangents of the primitives).

Some recognition systems do not solve these difficulties during the vectorization but rather delay solving these to higher level processing. However, we do not recommend delaying since it is seldom possible to qualify the resulting primitives with the corresponding raster data as the raster information is not kept after vectorization.

3 Proposed Solution

We propose introducing constraints for tangents of the primitives into the vectorization process (more precisely into the error fitting) as this will help solve some of the difficulties just mentioned. The main purpose of the error fitting is to connect adjacent raw primitives. In addition, we claim that the error fitting has to support constraints for tangents.

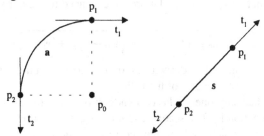

Fig. 2. On the left: tangents t_1 and t_2 defined from the endpoints p_1 and p_2 of the circular arc a. On the right: the same for a line segment s.

3.1 Constraints for Tangents

Fig. 2. shows how both a line segment and a circular arc have two tangents defined from their endpoints. We assign each tangent t_i a direction with an angle φ_i:

$$-\pi \leq \varphi_i < \pi \tag{1}$$

In general, a constraint for two tangents t_i und t_j can be defined by requiring a constant value c for the angle difference $\Delta\varphi = \varphi_i - \varphi_j$:

$$\Delta\varphi = c \tag{2}$$

Examples

- $c = 0$ implies that the tangents t_i and t_j are parallel (or collinear).
- $c = \pi/2$ implies that the two tangents are perpendicular.

3.2 Fitting Points

In general, optimizing a primitive means fitting it to a set of data points under the bounding condition of connecting adjacent primitives. If the primitive is a line segment, a line is fitted to this set of data points. The line is defined by the endpoints of the segment. Because there can be several primitives adjacent to one of these endpoints, we optimize the segment (line) by moving (fitting) the endpoints into an optimal position. The endpoints are the fitting parameters in the error fitting process so we call them the *fitting points*. If the primitive is a circular arc, a circle is fitted to the set of data points. To optimize the circle, we need an additional fitting point: the center of the circle. For the optimization process we use an iterative non linear least squares fitting [4]. So, we need an error function to compute the error distance of a data point to the fitted primitive. In the case of a segment this error function is based on the fitting parameters of the segment, i.e. on the Cartesian coordinates of the two endpoints. Given the endpoints $p_1(x_1, y_1)$ and $p_2(x_2, y_2)$ of the segment s. The corresponding error function F_s to compute the error of a data point $p(x, y)$ is:

$$F_s \equiv \frac{x(y_1 - y_2) + y(x_2 - x_1) + x_2 y_1 - x_1 y_2}{\sqrt{(y_1 - y_2)^2 + (x_2 - x_1)^2}} \tag{3}$$

To get an error function F_a for a circular arc we also need the center $p_0(x_0, y_0)$ of this arc:

$$F_a \equiv \sqrt{(x - x_0)^2 + (y - y_0)^2} - \sqrt{(x_1 - x_0)^2 + (y_1 - y_0)^2} \tag{4}$$

In addition, we also need the derivatives of the error functions to their fitting parameters, i.e. to the coordinates of the fitting points.

We establish that only one of the two endpoints of a circular arc is needed to define the error function F_a. The second endpoint is dependent on the first endpoint and on the center which define the corresponding circle. To express this dependency and to introduce the constraints for tangents into the error fitting process we use another fitting element, the *fitting vector*.

3.3 Fitting Vectors

A fitting vector v is a directed segment between any two fitting points. We describe v with its polar coordinates α and r. Fitting vectors have two important properties.

On the one side, they express dependencies among fitting points. A fitting point p_i can be declared dependent on another one by adding a fitting vector v to a fitting point p_j:

$$\begin{pmatrix} x_i \\ y_i \end{pmatrix} = \begin{pmatrix} x_j \\ y_j \end{pmatrix} + \begin{pmatrix} r*\cos(\alpha) \\ r*\sin(\alpha) \end{pmatrix} \tag{5}$$

If we assign a constant value to the angle α of vector v, point p_i lies on a line through p_j and if we assign a constant value to the radius r of vector v, point p_i lies on a circle with the center p_j.

On the other side, fitting vectors have a close relation to the tangents of the primitives:

- The vector between the two endpoints of a segment has the same direction as the two tangents of the segment.
- The vectors between the center of a circular arc and the two endpoints of this arc are perpendicular to the tangents from the endpoints.

So, we are able to define constraints for tangents by expressing relations between the angles of the fitting vectors. These relations are automatically introduced into the error fitting process with the vector dependencies on fitting points.

Optimizing a primitive now affects not only the endpoints of the primitive but all the fitting points which are directly or indirectly dependent on the endpoints. For this reason we need the derivatives of the error function F of a primitive to all the dependent fitting points of this primitive, and to all the fitting vectors between these fitting points.

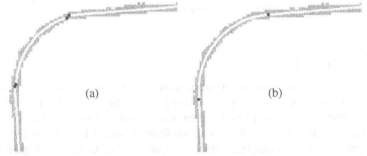

Fig. 3. Vectorization of a continuous curve: the segmentation finds the correct number of primitives and the desired type of each primitive (a). The error fitting optimizes the endpoints using constraints for tangents. The resulting primitives are shown in (b).

3.4 Fitting Process

Finally, the error fitting process consists of four main steps. In the first one, we examine the raw primitives generated by the segmentation and define constraints for tangents of these primitives. In a second step, the raw primitives and the constraints are translated into fitting points and fitting vectors. The third step is an iterative fitting process, the constraint fitting. And in the last step, we retranslate the fitted points and vectors into the final primitives (Fig. 3. + 4.).

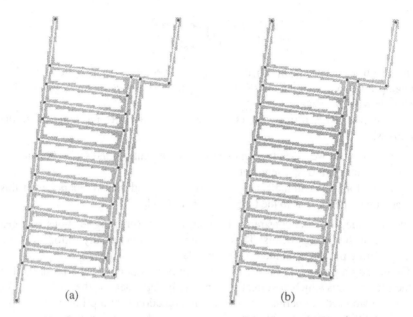

(a)   (b)

Fig. 4. The problem with short primitives: (a) shows the primitives connected without fitting with constraints. Alternatively, one can do a constraint fitting on the raw primitives; the resulting primitives are illustrated in (b).

4 System Overview

In many recognition systems, the vectorization is part of a sequential process. In contrast, in our system only the segmentation is part of this sequential process. The constraint fitting is separate from the main process and it is invoked several times by different modules (Fig. 5.). The constraint fitting is called for the first time after segmentation. Then, the constraint fitting is also suitable to support higher level modules. For instance, it is called during the construction of line objects (Fig. 6.). Further, the constraint fitting supports also the postprocessing by refitting the primitives after the manual correction or manipulation. The user may chose some primitives and impose a constraint on these primitives. This operation would be supported also by the constraint fitting module.

We conclude that not only does the constraint fitting play a major part in the vectorization, it is also used effectively by the higher recognition modules. And because the constraint fitting is independent of other components, it is also a reusable module for the construction of more general recognition systems.

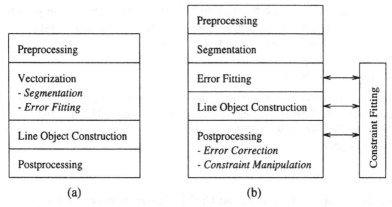

(a)   (b)

Fig. 5. In (a), the vectorization is one part in the sequential process of recognition. In contrast, (b) illustrates the constraint fitting as a part, separate from the main process, which is invoked at different times by the various modules.

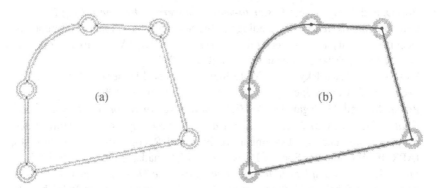

Fig. 6. In this figure we illustrate how the constraint fitting can be used during the line object construction, specifically during the construction of parcels in cadastral maps. (a) is the result after the first constraint fitting. After the primitives that correspond to boundary signs have been eliminated, the constraint fitting is called again (by the line object construction module). The result (the required boundaries) of this second constraint fitting is shown in (b).

5 Conclusions

In this paper we presented a way of satisfying constraints between primitives at the vectorization level. The motivation to introduce the "constraint fitting" into our vectorization came from our users who require us to deliver primitives "at right angles", primitives which are parallel, and so on.

The vectorization as presented is somewhat specialized. It should be used when the resulting primitives need to be of high quality. If this is not a requirement, the computation costs of this type of vectorization are too high. In addition, this vectorization is inappropriate for a raster-to-vector conversion of line primitives drawn with strokes (such as strokes used to represent Chinese characters) since the line width of the strokes is not constant.

6 Acknowledgments

We thank the financial support of the Swiss Federal Commission for the Advancement of Scientific Research (KWF), project 2540.1, and of the Aargauisches Elektrizitätswerk (AEW). AEW also provided us with the test data used in the project.

References

1. Dori D., Linag Y., Dowell J., and Chai I.: *Sparse-pixel recognition of primitives in engineering drawings.* Machine Vision and Applications, 6(2-3), pages 69-82, Spring-Summer 1993.
2. Grimson W. E. L.: *Object Recognition by Computer: The Role of Geometric Constraints.* MIT Press, Cambridge, MA, 1990.
3. Hori O. and Tanigawa S.: *Raster-to-vector Conversion by Line Fitting Based on Contours and Skeletons.* Proceedings of the Second International Conference on Document Analysis and Recognition, pages 353-358, IAPR, IEEE Computer Society Press, Tsukuba, Japan, October 20-22, 1993.
4. Press W. H., Teukolsky S. A., Vetterling W. T., and Flannery B. P.: *Numerical Recipes in C.* Cambridge University Press, second edition, 1992.
5. Röösli M. and Monagan G.: *A High-Quality Vectorization Combining Local Quality Measures and Global Constraints.* Proceedings of the Third International Conference on Document Analysis and Recognition, pages 243-248, IAPR, IEEE Computer Society Press, Montreal, Canada, August 14-16, 1995.
6. Simon J. C.: *A Complemental Approach to Feature Detection.* In J. C. Simon, editor, From Pixels to Features, pages 229-236, Elsevier Science Publishers B. V. (North-Holland), 1989.

Quantitative Measurement of the Performance of Raster-to-Vector Conversion Algorithms

Osamu Hori and David S. Doermann

Document Processing Group
Center for Automation Research
University of Maryland
College Park, MD 20742, USA

Abstract. This paper presents a methodology for measuring the performance of application-specific raster-to-vector conversion algorithms. In designing and building image analysis systems, comparison of several algorithms is often required. Unfortunately, many methods of comparison do not give quantitative performance measurements, but rather qualitative, and often subjective, evaluations. Our key observation is that there is a need for domain, or task-dependent evaluation of the output. By specifying the input data in the same parameter space as the intended output of the system, we are able to evaluate the quality of the output and how well it conforms to the intended representation. We provide a set of basic metrics, but we emphasize that in general, such metrics may be task-specific. In this paper, the performance of three approaches to raster-to-vector conversion - *thinning, medial line finding,* and *line fitting* - are compared using this methodology.

1   Introduction

Raster-to-vector conversion is an important task in graphics recognition. Many papers proposing raster-to-vector conversion algorithms continue to appear in the research literature. In designing, building, and improving image analysis systems, comparison of several algorithms is often required. Unfortunately, many methods of comparison do not give quantitative performance measurements, but rather qualitative, and often subjective, evaluations.

Haralick [2] proposes a philosophy and framework for performance characterization in image analysis in which an ideal input is specified. A random perturbation model specifies how the data is perturbed and how perturbed images are generated. A criterion function which quantitatively measures the difference between the ideal input and the calculated output is given, and is then used to compare algorithms. Jaisimha, Haralick and Dori [5] characterize thinning algorithms based on Haralick's framework and provide objective measurements which are domain independent. In related work, Lee et al. [7] give measurements for thinning algorithms relating to the accuracy of OCR systems performance.

The measurement of the performance of raster-to-vector conversion algorithms has not yet been treated in the research literature, but it remains an important component in studying how to improve existing algorithms. In this paper we present a methodology for measuring the performance of raster-to-vector conversion algorithms by instantiating Haralick's framework and extending it in an application-dependent manner.

In Section 2, our approach is described as a set of modifications and extensions to Haralick's original framework. In Section 3, we describe the basic components of an evaluation system which would be applicable to mechanical engineering drawings. In Section 4, we demonstrate our system on *thinning*, *medial line finding*, and *line fitting* raster-to-vector conversion algorithms. Finally, discussions and conclusions are presented in Section 5 and 6, respectively.

2 Approach

Our research suggests that there do not exist generic measurements which are independent of the intended applications. For example, thinning is often used as a pre-processing step for optical character recognition (OCR). Individual OCR classifiers, however, may require different features. If an OCR classifier uses the locations of crossing points, corners, T-junctions, and end points, the thinning algorithm must precisely preserve these feature points. If an OCR classifier uses only the directions and lengths of character strokes, however, the locations of feature points are less important. When an algorithm is evaluated, specifying the requirements of the application is necessary.

The key observation in our approach is that there is a need for domain, or task-dependent evaluation of the output. By specifying the input data in the same parameter space as the intended output of the system, we are able to evaluate the performance characteristics of the output, and how well it conforms to the intended representation. Although a set of basic metrics is given here, in general, such metrics may be task-specific.

Figure 1 shows the framework of the proposed evaluation method. The first step is to specify the ideal input data. The input data that are used to evaluate the performance, must contain patterns meeting the requirements of an application, and be representative of the real data. In most cases, the input to the analysis system is an image. If the ideal input is specified as a set of features, an approach to synthesizing the image must also be provided. Second, a perturbation model should provide a way to perturb the image generated from the ideal input data, so that we obtain a set of data which is representative of the domain of the application. It may, for example, simulate the effects of imaging such as binarization, blur, skew, and so forth. Third, an application-dependent procedure to calculate the deviation between the ideal input data and the output data is required. For example, a mid-line in a line image, along with its thickness, is required as the ideal representation in many mechanical engineering drawing applications. For electrical diagrams, however, a precise location and thickness are not always required, because the connections between symbols such as diodes and transistors are the most important. In the first case, the errors in line thickness and position should be measured; in the second case, the deviation can be measured as errors in recognizing connections. The matching process between the ideal input data and the output data should consider the application-dependent requirements.

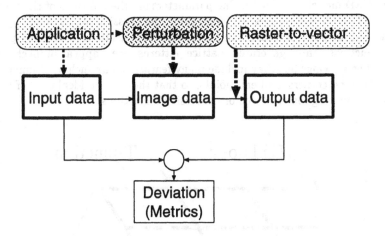

Fig. 1. The framework of the proposed method.

3 Instantiation of the Method

In this section, we describe in detail a simple system to evaluate the performance of a raster-to-vector conversion algorithm based on Haralick's framework. The features used are not comprehensive, but do provide a demonstration of the approach.

3.1 Specifying the Application Output Feature Set

One of main goals of raster-to-vector conversion is to convert from paper-based line drawings into reusable CAD-formatted numeric data. A second use of raster-to-vector conversion is pre-processing to recognize symbols in line drawings. We have chosen to illustrate our approach on mechanical engineering drawings.

The first step is specifying the application output features; we base our analysis on a CAD representation. A line in a CAD representation is described by the locations of two end points, the line thickness, and possibly a line type such as broken, dotted, or solid. We claim, however, that these are not enough to evaluate the performance of raster-to-vector algorithms. T-junctions and crossing points, which often appear as feature points in line drawings, are also needed to evaluate the performance, because they significantly influence line drawing quality.

3.2 Generating Artificial CAD Models

The ideal inputs are lattice patterns with varied crossing angles, including the identification of T-junctions, crossing points, corners, and end points, which are used in CAD models (Figure 2). The parameters are the positions of the feature points and the line thickness. By controlling the crossing angles made by the intersections of parts of straight lines, various configurations of crossing points and T-junctions can be generated. Lattice patterns rarely appear in mechanical engineering drawings but they include many feature points such as T-junctions, crossing points, corners, and end points, so that they are an ideal type of input data for evaluating raster-to-vector conversion algorithms.

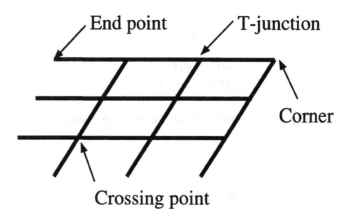

Fig. 2. An example of a lattice pattern.

3.3 Generating Synthetic Image Data from CAD Models

The ideal input features are mapped in the image space in a straightforward manner. The X-windows library is used to synthesize line images with different line thicknesses and the numeric CAD data are converted into bitmap image data. A perturbation model is important to simulate real world conditions; it is applied to the synthetic image to generate degraded images.

3.4 Processing the Generated Images and Evaluating the Results

The outputs of processing the synthetic (and resulting degraded) images are matched to ideal CAD models and the discrepancies between the input and output data are calculated. The numbers of missing and matching feature points and lines, and the accuracies of their locations provide quantitative measures.

Evaluation of Feature Points Each input feature point is matched to the nearest output feature point within a circle. The numbers of missing and matching feature points are counted and divided by the total number of input feature points to provide "point matching ratios". The radius of the circle is defined as $t = w/2 + e$, where w is the line thickness and e is an allowable margin of error (Figure 3). The Euclidean distance between the matching feature point and the corresponding input feature point is calculated as a deviation. The feature point types are T-junctions, crossing points, corners, and end points. The matching ratios for each feature point type reflect the performance characteristics of the raster-to-vector conversion algorithm.

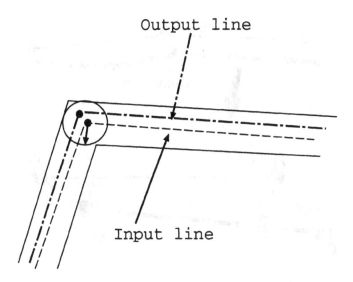

Fig. 3. An example of matching points.

Evaluation of Lines The input lines are also matched to the output lines, and the total length of the matching lines is divided by the total length of the corresponding input lines, resulting in a "line matching ratio". The average Euclidean distance between the matching lines and the corresponding input lines is calculated as the line deviation. The number of matching lines corresponding to a given input line is also counted.

An output line corresponding to one input line might be segmented into several fragments by the raster-to-vector conversion process, but might still be continuous. A matching procedure is used as shown in Figure 4 to resolve this ambiguity. First, one of the input lines is selected and the neighboring output lines are retrieved. Each set is checked to see if the input line corresponds to the retrieved output line or not by calculating the distance between them. For example, let a and b be the end points of the input line, and c and d the end points of the output line. The lengths of the perpendicular lines from each point to the opposite line are measured. If a perpendicular line does not exist, the distance between the point and the nearest end point of the opposite line is used. If the condition "$min(da, dc) < t$ and $min(db, dd) < t$" is satisfied, where t is the radius of the circle used in the point evaluation, the retrieved output line matches the input line; otherwise, it does not match. Let a' and b' be the end points of the matching output line. The distance between a' and b' each match is summed up and divided by the total length of the input lines to compute the line matching ratio.

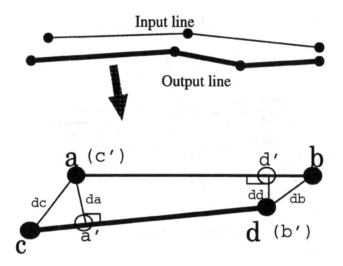

Fig. 4. How to match the input and output lines.

4 Experiments

4.1 Sample Algorithms

In this paper, the performance of three approaches to raster-to-vector conversion - *thinning* [8], *medial line finding* [6], and *line fitting* [4] - are compared using this methodology. The thinning approach uses an iterative boundary erosion process that removes pixels until a unit-wide pixel chain remains, and then converts the chain into vectors. The extracted vectors are not processed further so that they have only end points, but T-junctions or crossing points. T-junctions and crossing points in the original image appear as "junctions", where several end points meet at a single point in the processed image.

The medial line finding approach extracts image contours as pixel chains which are approximated by polygons. The mid-points of two parallel contours, given by the mid-points of perpendicular lines projecting one side on the other, define the vectors. Pairs of parallel contour lines are often missing at T-junctions, crossing points, and corners, so that mid-points are not extracted completely using this approach. End point joining and vector merging processes are necessary to identify the T-junctions, crossing points, and corners; a post-processing stage is added for that purpose. The line fitting approach makes use of both parallel contours and thin lines. The best fitting line among all candidates is chosen as the resulting vector. When pixel chains are approximated by polygons, the maximum deviation of the extracted vectors from the pixel chains in the contours or thinned images is set to 1 pixel.

4.2 Sample Data

Two types of lattice patterns were prepared and processed by the three algorithms. One is composed of horizontal and vertical lines, which cross each other perpendicularly. The other is composed of horizontal and diagonal lines meeting at fixed angles. The former, "rectangular lattices" had three different line thicknesses (3-pixel, 5-pixel, and 7-pixel), and the latter, "parallelogram lattice patterns" had three different angles (30 degrees, 38 degrees, and 45 degrees). The left-hand image in Figure 5 shows an example of the rectangular lattice pattern image and the right-hand image shows an example of the parallelogram lattice pattern image.

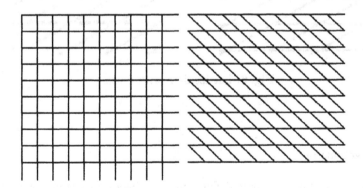

Fig. 5. Rectangular and parallelogram lattice data.

The images, whose sizes are 800 × 800 pixel, are generated from these lattice patterns, which are composed of 20 horizontal and 20 vertical or diagonal lines, using the X-windows library. They are then distorted by the perturbation algorithm [1] which simulates the effect of photocopying binary images. This perturbation model is based on the distance transform [3] of the ground truth data and on some morphological post-processing. The probability of a pixel changing from its ideal value as a function of the distance of that pixel from the boundary of an image is modeled. In order to generate the noisy images, the algorithm is used as follows:

- Each pixel in the foreground is inverted with a probability of
  $P(0|foreground, d) = \gamma + \alpha 0 \times \exp^{-\alpha d^2}$
  where d is the distance of the foreground pixel from the background.
- Each pixel in the background is inverted with a probability of
  $P(1|background, d) = \gamma + \beta 0 \times \exp^{-\beta d^2}$
  where d is the distance of the background pixel from the foreground.
- A morphological closing with a disk of diameter δ is performed.

In our experiment, we set the parameters as follows: $\gamma = 0.0$, $\alpha 0 = 1.0$, $\alpha = 1.0$, $\beta 0 = 1.0$, $\beta = 1.0$, and $\delta = 3$. Figure 6 shows a portion of the original parallelogram lattice pattern image and the resulting distorted parallelogram lattice pattern image.

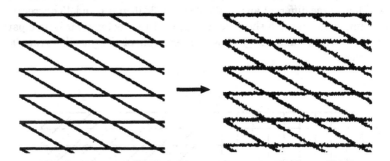

Fig. 6. Lattice data with noise.

4.3 Evaluation

The purpose of this paper is to define a methodology for obtaining quantitative measurements for evaluating the performance characteristics of the algorithms, and not necessarily to choose the best algorithm among them. The meaning of "the best algorithm" is defined by the intended application. Similarly, how an algorithm is implemented also influences its performance. Our results show that the quantitative measurements conform to the performance characteristics. The parameters of the three algorithms were optimized to make the best performance for rectangular lattice data with 3-pixel line thickness.

Evaluation for Rectangular Lattice Data Table 1 shows the point matching ratios for the rectangular lattice. For T-junctions, the ratio decreases as the line becomes thicker in the case of the medial line finding algorithm. This is because the end point joining algorithm used as a post-processing step is sensitive to the parameters and needs to be adjusted to the line thickness.

The medial line finding algorithm uses only line image contours so that it is influenced by noise on the contours. T-junctions and crossing points are extracted at 100% matching ratio from 3-pixel-thickness lines without noise. However, in the noisy image, the matching ratio goes down to 82% and 62% for the 5- and 7-pixel cases respectively. This means that the end point joining algorithm is not robust to noise.

In terms of the location deviations, there are no differences among the three algorithms in the rectangular lattice experiments. The count ratio shows how many output lines correspond to an input line. This ratio is closely related to the point matching ratio of the crossing points. When the point matching ratio of the crossing points is low, output lines are segmented into several shorter lines.

Table 1. Point evaluation for rectangular lattice data. Thi, Med, and Fit represent the thinning, medial line finding, and line fitting algorithms respectively. The postfix numbers indicate line thickness. For example 3 corresponds to a 3-pixel line thickness. EN, TJ, CR, and CO stand for end point, T-junction, crossing point, and corner respectively. The percentages are point matching ratios and the values in parentheses are deviations which are defined as Euclidean distances between the original and extracted points.

	Original				Noise			
	EN	TJ	CR	CO	EN	TJ	CR	CO
Thi3	100%(0.5)	0%(0.0)	0%(0.0)	100%(0.8)	98%(0.9)	0%(0.0)	0%(0.0)	68%(0.7)
Thi5	100%(1.0)	0%(0.0)	0%(0.0)	100%(1.0)	98%(0.9)	0%(0.0)	0%(0.0)	73%(0.9)
Thi7	100%(1.0)	0%(0.0)	0%(0.0)	100%(1.8)	98%(1.0)	0%(0.0)	0%(0.0)	90%(1.4)
Med3	100%(1.0)	100%(0.7)	100%(0.6)	100%(0.3)	100%(1.1)	82%(0.4)	62%(0.5)	100%(1.3)
Med5	100%(1.0)	97%(0.7)	100%(0.7)	100%(0.3)	100%(1.1)	71%(0.6)	47%(0.4)	95%(1.1)
Med7	100%(1.0)	50%(0.6)	100%(1.4)	100%(0.4)	98%(1.1)	71%(1.0)	49%(0.7)	90%(1.1)
Fit3	100%(0.5)	100%(0.2)	100%(0.7)	100%(0.8)	100%(1.0)	100%(1.1)	97%(1.1)	98%(1.1)
Fit5	100%(1.0)	100%(0.5)	100%(0.9)	100%(1.0)	100%(0.9)	100%(0.6)	99%(0.8)	95%(1.5)
Fit7	100%(1.0)	100%(0.7)	100%(1.2)	100%(1.8)	100%(1.0)	100%(1.1)	97%(1.2)	98%(1.6)

Evaluation for Parallelogram Lattice Data Tables 3 and 4 show the results of processing the parallelogram lattice data. There is not much point deviation in the rectangular lattice data, but there is point deviation in the parallelogram lattice data, especially in terms of T-junctions extracted by the line fitting algorithm. On the other hand, the medial line finding algorithm gives good performance in terms of the point deviations. The fitting method uses thinning

Table 2. Line evaluation for rectangular lattice data. The length ratio is the total length of the matching lines divided by the total length of the corresponding input lines. The deviation is the average Euclidean distance between the matching output lines and the corresponding input lines. The count ratio is the number of matching output lines corresponding to an input line.

	Original			Noise		
	Length ratio	Deviation	Count ratio	Length ratio	Deviation	Count ratio
Thi3	100.9%	0.08	20.0	102.4%	1.16	36.0
Thi5	100.9%	0.08	20.0	102.3%	0.94	34.4
Thi7	100.8%	0.56	20.0	102.9%	1.15	36.3
Med3	100.0%	0.13	1.0	99.3%	0.64	7.2
Med5	100.0%	0.14	1.0	98.9%	0.68	9.9
Med7	99.6%	0.08	1.0	98.5%	0.62	9.7
Fit3	100.0%	0.13	1.0	100.0%	1.57	1.5
Fit5	100.0%	0.26	1.0	100.1%	0.98	1.1
Fit7	100.0%	0.35	1.0	100.2%	1.26	1.3

information to get the locations of T-junctions so that the locations have some errors generated during the thinning process. The point matching ratios of T-junctions and crossing points are good in the line fitting algorithm in comparison with those in the other algorithms.

When two lines cross at an acute angle, T-junctions and crossing points are significantly distorted by the thinning process. The medial line finding algorithm has a problem in the end point joining process because the distances between the end points at intersections are so long that the end point joining algorithm does not operate properly (Figure 7).

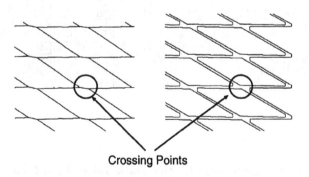

Crossing Points

Fig. 7. Distorted thinned lines and contours.

Table 3. Point evaluation for the parallelogram lattice data. For an explanation of the notation, see Table 1. The postfix numbers represent crossing angles, for example, 30 stands for a 30 degree angle.

	Original			Noise		
	TJ	CR	CO	TJ	CR	CO
Thi45	0%(0.0)	0%(0.0)	50%(1.5)	0%(0.0)	0%(0.0)	16%(0.5)
Thi38	0%(0.0)	0%(0.0)	50%(2.0)	0%(0.0)	0%(0.0)	5%(0.2)
Thi30	0%(0.0)	0%(0.0)	0%(0.0)	0%(0.0)	0%(0.0)	3%(0.1)
Med45	95%(0.2)	100%(0.8)	100%(1.0)	81%(0.9)	17%(0.1)	95%(1.1)
Med38	86%(0.6)	74%(0.7)	100%(0.6)	76%(1.0)	2%(0.0)	94%(1.3)
Med30	58%(0.9)	0%(0.0)	100%(0.6)	50%(0.9)	0%(0.0)	82%(1.4)
Fit45	100%(1.8)	100%(1.7)	100%(2.0)	97%(3.0)	90%(1.7)	95%(1.4)
Fit38	100%(3.0)	100%(1.6)	100%(2.0)	96%(3.7)	90%(1.9)	97%(1.0)
Fit30	100%(4.0)	100%(1.9)	100%(1.0)	90%(3.5)	83%(2.4)	66%(0.6)

Table 4. Line evaluation for parallelogram lattice data. The explanation of the notation, see Table 2.

	Original			Noise		
	Length ratio	Deviation	Count ratio	Length ratio	Deviation	Count ratio
Thi45	103.4%	0.57	25.7	109.0%	1.79	39.8
Thi38	102.3%	0.68	21.8	112.3%	2.14	39.5
Thi30	102.2%	0.73	18.1	116.5%	2.43	40.8
Med45	99.8%	0.40	1.0	102.7%	1.08	10.4
Med38	100.2%	0.79	2.2	105.8%	1.14	13.9
Med30	103.4%	0.72	6.7	104.8%	1.46	14.7
Fit45	99.7%	0.61	1.0	99.6%	1.06	1.7
Fit38	99.7%	0.60	1.0	99.7%	1.26	1.9
Fit30	99.7%	0.41	1.0	98.4%	1.34	2.2

5 Discussion

The above evaluation makes some shortcomings of the algorithms clear. The medial line finding approach requires a robust end point joining algorithm that should be adaptive to image line thickness and the line fitting approach needs to adjust the positions of lines, distorted by the thinning process, using contour information.

Individual measurements play important roles in specifying the performance characteristics of an algorithm. Other measurements of performance, such as sensitivity to image rotation or tuning parameters, should also be examined.

The results of experiments can then be used to adjust the algorithm parameters for a specific application.

The proposed methodology gives multi-dimensional evaluations, so that the most suitable algorithm for a specific application can be chosen by considering the evaluations from a practical, quantitative point of view. For example, in situations where large quantities of data must be handled, it is impossible to adjust the parameters of an algorithm for each data set. The algorithm which is best with fixed parameters should be selected in such a case.

6   Conclusion

In the graphics domain, quantitative evaluation algorithms for raster-to-vector conversion have not been fully explored. Our approach, which recognizes the need for application-specific evaluation, instantiates Haralick's framework for system evaluation and a step toward robust evaluation.

This methodology can be easily extended to the performance evaluation of the arc, circle, string, and symbol extraction algorithms that are often used in analyzing engineering line drawings.

References

1. Kanungo, T., Haralick R. M., Phillips, I.: Global and Local Degradation Models. Proceedings of ICDAR'93 (1993) 730–732
2. Haralick R. M.: Performance Characterization in Image Analysis: Thinning, a Case in Point. Pattern Recognition Letters 13 (1992) 5–12
3. Haralick, R. M., Shapiro, L. G.: Machine Vision Addison-Wesley, Reading, Mass. (1992)
4. Hori, O., Tanigawa, S.: Raster-to-Vector Conversion by Line Fitting Based on Contours and Skeletons. in Proceedings of ICDAR'93 (1993) 353–358
5. Jaisimha, M. Y., Haralick, R. M., Dori, D.: A Methodology for the Characterization of the Performance of Thinning Algorithms. Proceedings of ICDAR'93 (1993) 282–286
6. Jimenez, J., Navalon, J. L.: Some Experiments in Image Vectorization. IBM J. Res. Develop. 26 (1985) 724–734
7. Lee, L., Lam, L., Suen, C. Y.: Performance Evaluation of Skeletonization Algorithms for Document Image Precessing. Proceedings of ICDAR'91 (1991) 260–271
8. Ramer, U. E.: An Interactive Procedure for Polygonal Approximation. CGIP, Vol. 1 (1972) 244–256

Form Item Extraction Based on Line Searching

E. Turolla, Y. Belaid, A. Belaid

CRIN-CNRS, Bât. Loria, BP 239,
54506 Vandoeuvre-lès-Nancy Cedex, France
Tel: (33) 83.59.20.83 - Fax: (33) 83.41.30.79
E-mail: {turolla, ybelaid, abelaid}@loria.fr

Abstract. This paper presents an item searching method which has been applied to various kinds of forms. This approach is based on line detection through the Hough transform. After obtaining the straight lines, Hough directions are used to detect the real segments in the image. Segments can correspond either to continuous line, or to black parts of dashed or dotted lines. So, the segments are grouped together and classified between both adjacent line crossing points. Items are located by searching the minimum cycles of the graph constructed from the line intersection points. The last step consists of verifying the line classes based on the homogeneity hypothesis of item sides.

This method was applied to French Tax forms and tables coming from scientific publications. The experimental results have demonstrated the robustness and the reliability of such an approach to various forms with different types of item delimiters.

Keywords: Form Analysis, Item Extraction, Line Searching, Hough Transform.

1 Introduction

Forms and tables are often structured into items clearly delimited by horizontal and vertical lines. So it is natural to base the form analysis methods on line searching. In spite of noise and defects, lines constitute reliable frontiers between items, more accurately than white streams. Furthermore, damage suffered by lines can be "easily" overcome by a global analysis.

It could be thought that line extraction is easy because lines in forms are only straight and limited to two directions. However, some of their features (length and thickness) make their extraction difficult. Moreover, the line type can be diverse : continuous, dashed, dotted or composed of brackets, that leads to the use of more specific methods.

Existing literature mentions few form analysis systems based on line searching, because such applications usually deal with filled pre-printed forms whose physical structure is frozen. Thus, items are directly located by the subtraction between the filled and the blank forms [1]. In the case of changing structure, methods based on line searching should be used. Yan *et al.* [2] use information such as the first point and the line length to locate the lines within telephone bills. Wong *et al.* [3] base their approach on the detection and examination of connected components extracted by the smoothing algorithm RLSA; the main features considered are height and mean run length of black pixels. Another approach, used by Watanabe [4] and Taylor [5], is to search for lines in locating corners and line crossings by applying a set of specific masks.

Lines provide useful topological features for document analysis and skew estimation. For example, Lam *et al.* [6] detect lines and skew angle, from contours of connected components; the consecutive chaincodes on a contour that have similar direction represent a line whose direction gives the skew angle estimation. Another class of technics use a more global approach to extract lines, such as those based on projection histogram or on Hough transform [7].

This paper proposes an experimental structure analysis method for table-form documents based on line searching. This approach uses Hough transform for its reliability and accuracy. It was applied on French tax forms. This type of form is not mapped regularly, lines are skewed and the item type is diverse. The particularity of this method is that line searching is based on item searching, so reliable criteria is used on items frontiers to classify the lines in the searching categories : continuous, dashed and dotted lines. Results obtained on French tax forms and tables are discussed at the end of this paper.

2 Form Description

A French tax form is a set of pages; each one is composed of one or more tables with different formats. Fig. 1 shows a page from this form. The logical structure (meaning and relative positions of the items) is unvarying whereas the physical structure may change every year. In the same form, the dimensions and the separators of the same item can differ. So, the position of each item is not predictable. It is no use trying to recognize the form page number located at its top right corner, because it gives no information about its layout.

Lines are very diverse; except the continuous lines easily detected by a classical method, the other delimiters are more "exotic". Vertical alignments of little brackets, or big brackets can be found as item delimiters. They are in front of several lines and close items. In some cases, lines are also delimiters of black items (black background), and some vertices are rounded. Finally, these kind of forms are often noisy and skewed, moreover, items are filled with characters which may overlap the lines.

In order to test our line searching system, we have also studied the lines in the context of tables from scientific publications. Indeed, these tables are a minefield of diverse lines such as dotted, dashed and double.

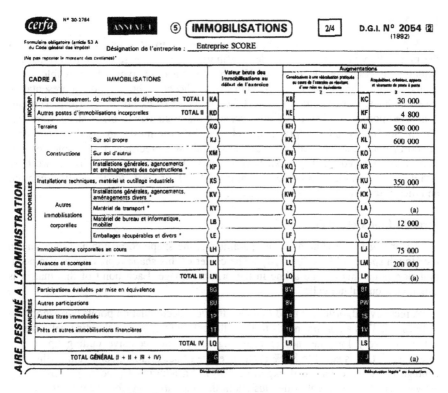

Fig. 1. An extract of a French tax form.

3 Line Searching

Line searching is elaborated into three steps : the first one leads with straight line detection, the second one is reserved for the extraction of the effective segments within the image; the last step consists of finding the item delimiters and classifying them.

3 .1 Straight Line Detection

The lines are diverse and the documents are often noisy and skewed. So, a global approach (the Hough transform) is used to first detect the straight line before searching the effective black segments.

The Hough parameter space (α, ρ) [8] have been chosen. α is the slope of the line with respect to the x-axis, and ρ is the distance between the line and the origin of the cartesian space (the top left vertex of the image). Two bidimensional matrices are used to represent the Hough space, one for horizontal lines and the other for vertical lines. These matrices are filled with pixel voting, described in the following section.

Pixel Vote

A pixel votes for every straight line it can belong to. But, to avoid treating all of the pixels, they are filtered. The idea of filtering is to increase the Hough transform velocity, to decrease the noise in Hough space and to avoid the possible over-segmentation of thick lines. Pixels retained are chosen to be the most representative pixels of lines. First, we tried to select the middle of black horizontal (resp. vertical) connected points. But it was a failure because the boudaries of black strips were forgotten. So, finally, contour pixels are selected. Specific masks are used to keep only the horizontal (resp. vertical) contours. Masks allow a precise and a flexible description of the selected pixel context. Several masks have been tested. Finally, we have retained these simple mono-dimensional ones: "100" or "001" for the vertical contour and their transposition for the horizontal ones.

Straight Line Detection

According to the Hough chosen parameters, the shape corresponding to a line in the Hough space is like a "butterfly". The center of this shape represents a vote accumulation point that can be a line in the original image. Each point of this matrix is called an accumulator cell.

A recursive algorithm is used to detect the vote accumulation points corresponding to lines [9]. This algorithm needs to filter the cell before, by thresholding cells whose values are very low, i.e. not representing lines. The recursive cutting up uses a threshold T, that is equal to 83 % of the average of accumulator cells values (this percentage has been determined experimentally). An additional process groups all the adjacent clusters or neighboring clusters in the line direction (for instance, the search direction for horizontal lines is about 45°) and filters the remaining low clusters.

3 .2 Segment Searching

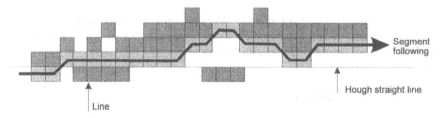

Fig. 2. Segment searching.

The searching of black segments consists in following each Hough line on the image and merging together into one segment, the black consecutive points which are the nearest to the Hough line. Fig. 2 is an example of segment following. A character that touches a line does not disturb the searching. But the noise that exists on the Hough line (e.g. parts of character), is taken into account and will have to be eliminated during the following step.

3 .3 Item Searching

Line Graph Construction

From found straight lines, a graph is constructed. Its nodes correspond to the intersections of the lines. Arcs should represent item delimiters. So, the items are the minimum cycles of the graph. The detection of the lines as item delimiters, is based on this graph description. Each node contains two things : the location within the image of the intersection point of the ith horizontal line and the jth vertical line and the neighbors nodes which are linked to this point.

A dashed or dotted line has several segments whereas a continuous line contains one segment. The arc construction consists in trying to link segments that exist between two graph adjacent nodes, according to criteria that depends on the type of line searched.

The arc searching algorithm is described in Fig. 3. The main function *FindArcs* looks for a dotted line between two consecutive nodes M_i and M_{i+1}; if it fails, then it looks for a continuous line, and if it fails again, then in the end it searches for a dashed line. A dotted line is looked for before a dashed line because a dotted line is quite similar to a dashed line whose segment lengths are very small. A continuous line needs to be found before a dashed line because a continuous line is quite similar to a dashed line whose gaps between segments are very small. The set $S_i^{\perp}$ corresponds to the parts of characters that are lined up with a Hough line or the segments that are touched by characters. It is so difficult to distinguish between these two cases that dashed or dotted lines whose set $S_i^{\perp}$ are not too important, are accepted. The threshold D_n is the maximum distance between a line and its crossing point (for dashed or dotted line, this point is not always inside a segment); d_{DL} is the maximum length of a dotted segment, E_{DL} is the maximum rate of long segments inside a dotted line; G_{DL} is the maximum gap between two dotted or dashed segments and G_{CL} is the maximum gap inside a continuous line. These thresholds depend widely on the image resolution. Nowadays, they are determined experimentally.

As soon as all the arcs have been determined, those that cannot be item delimiters are suppressed, by removing recursively the arcs that lead to nodes whose degree is one.

During the arc construction, an attribute corresponding to the line type (continuous, dashed and dotted) was determined for each arc. As an item delimiter can be composed of several collinear arcs and one delimiter should have only one line type, so the attributes of the arcs joining two item vertices are checked. Thanks to the last step, the item vertices correspond to the nodes that have both a vertical and a horizontal arcs. If an arc joining two item vertices is classified as "dashed", then it is verified whether the whole line between them is "dashed" (the gap between all the segments must be constant and the number of their different lengths must be small); if it fails, then the line is classified as "continuous". In the other cases, the type of the line is the type whose total segment length is the greatest.

```
Function FindArcs
begin
    for each neighboring node Mi and Mi+1, do :
        Calculate the set Si of the segments that exist between Mi and Mi+1 ;
        if (distance (Si, Mi) < Dn) and (distance(Si, Mi+1) < Dn) then
            if IsDottedLine(Si) then TypeArc = DottedLine;
                else if IsContinuousLine(Si) then TypeArc = ContinuousLine;
                    else if IsDashedLine(Si) then TypeArc = DashedLine;
                        else TypeArc = NoLine;
                    endif
                endif
            endif
            if TypeArc <> NoLine then it exists an arc whose type is TypeArc, between Mi and Mi+1;
            endif
        endif
    end do
end

function IsDottedLine (Si) return boolean
begin
    calculate Si² = {segment s / s ∈ Si and length(s) > dDL };
    calculate Si⊥ = {segment s / s ∈ Si and s has an orthogonal prolongation} (in order to not be
            disturbed by the eventual lines that cross Mi and Mi+1, the orthogonal prolongations are not
            searched into the areas that are near to Mi or Mi+1);
    if gap between two neighboring segments of Si < GDL and total length of the segments of Si²
            < EDL and card(Si⊥) << card(Si) then return true;
        else return false;
    endif;
end;

function IsContinuousLine(Si) return boolean
begin
    if gap between two neighboring segments of Si < GCL then return true;
        else return false;
    endif;
end;

function IsDashedLine (Si) return boolean
begin
    calculate Si⊥ = {segment s / s ∈ Si and s has an orthogonal prolongation} (in order to not be
            disturbed by the eventual lines that cross Mi and Mi+1, the orthogonal prolongations are not
            searched into the areas that are near to Mi or Mi+1);
    if gap between two neighboring segments of Si < GDL and card(Si⊥) << card(Si) then return true;
        else return false;
    endif;
end;
```

Fig. 3. Arc searching algorithm.

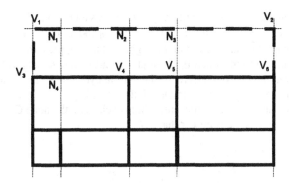

Fig. 4. Error during the arc classification.

For instance, in Fig. 4, V_1, V_2,...V_6 are item vertices. The item delimiter V_1V_2 consists of 4 arcs : V_1N_1, N_1N_2, N_2N_3 and N_3V_2. During arc classification, the unique segment between V_1 and N_1 is so small and so near V_1 and N_1, that the arc has been wrongly classified as "continuous", whereas the others arcs of V_1V_2 have been classified as "dashed"; this mistake is detected and corrected during this stage.

Item Searching

A minimum cycle is a series of consecutive and connected nodes where the first and the last are the same, and where each node has only two neighbors. For cycle searching, the idea is to start the search from one node chosen here as the top left vertex. Then, the algorithm tries to find the minimum clockwise path that can close the cycle.

Let E_{ca} be the set of top left vertices $\{(i,j)\}$ (a top left vertex is a node which has a horizontal arc on the left and a vertical arc at the bottom). E_{ca} is totally ordered according to the relation $\angle$:

$$(i_1, j_1) \angle (i_2, j_2) \Leftrightarrow (i_1 \leq i_2) \vee ((i_1 = i_2) \wedge (j_1 < j_2))$$

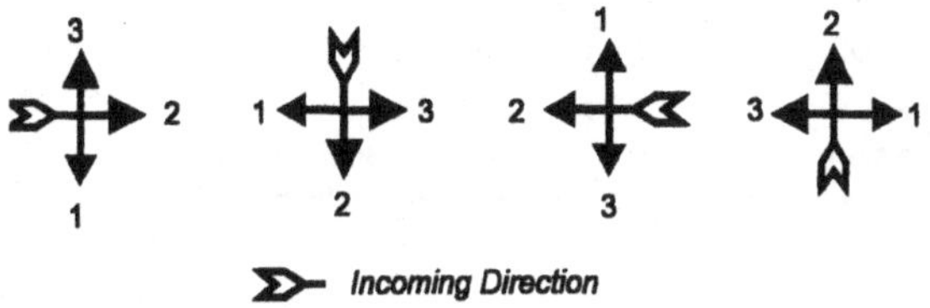

Fig. 5. Priority of the direction, according to the incoming direction.

The search starts from the node $N_0 = Min_{\angle}(E_{ca})$ and proceeds to the right. When a node N is reached from the direction d, the search proceeds towards the direction where an arc exists and whose priority is the lower (see Fig. 5).A partial failure is met when the current node is prior to the cycle or above N_0, then the search backtracks. If it backtracks till N_0, then it is a complete failure and it chooses a new start node. To avoid finding the same cycle several times, the start node N_0 and all the nodes that are reached from bottom ($\uparrow$) and are left on the right ($\rightarrow$) are suppressed from E_{ca}. All the cycles of the graph are found when E_{ca} is empty. To accelerate the algorithm, the nodes that have led to a partial failure for the current cycle are stored; if any of these nodes is reached again, then it is a complete failure. On Fig. 6, there are four top left vertices C_1, C_2, C_3 and C_4; the system finds three cycles and the node C_2 leads to a failure because its next node is higher than C_2.

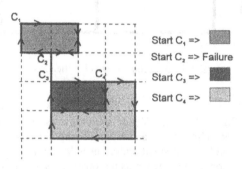

Fig. 6. Examples of cycle searching.

4 Experiments And Discussion

	OUI		NON		ABSTENTION		TOTAL
	n_i	np(oui)	n_i	np(non)	n_i	np(abs)	
Neuilly	15000	14286	16000	17143	9000	8571	40000
Corbeil	6000	7143	10000	8571	4000	4286	20000
Mazamet	4000	3571	4000	4286	2000	2143	10000
	25000	25000	30000	30000	15000	15000	70000

▨ Continuous line   ▨ Dashed line

Fig. 7. Classification of the item delimiter of a table.

This item searching method has been applied on 49 images of French tax forms and 114 tables, scanned at a resolution of 300 dpi. The treatment of one page of a French tax form takes about 30 secunds on a SUN SPARC 3.

Fig. 7 shows the result of table lines classification. Table 1 and Table 2 summarize the performances of line detection and classification. Lines considered in these tables correspond to the line joining two adjacent vertices. The edges of black strips, the brackets and the double lines, are recognized as continuous lines. Merged characters do not prevent the lines beeing found. However, if they are numerous, they disturb the classification of dashed and dotted lines. Continuous and dashed lines are well classified, but the dotted lines are too often classified as dashed. This will be improved by an additional test on the dot shape. It can be observed that the lines whose total length is smaller than 8 % of the longest total line in the image, are missing; the Hough transform does not manage to detect their straight lines because the threshold T used in Hough space analysis, depends on the mean length line.

Line type	Quality	Total number	Well classified lines	Bad classified lines	Missed lines
Continuous	Unskewed	14 188	92.55 %	0.01 %	7.44 %
	Skewed	4 003	87.96 %	0.04 %	12 %
	Merged characters	41	92.68 %	0.01 %	7.31 %
	Total	18 232	90.08 %	0.01 %	9.91 %
Boundaries of black strips	Unskewed	1 191	98.66 %	0.01 %	9.91 %
	Skewed	420	97.38 %	0.02 %	2.6 %
	Total	1 611	98.38 %	0.01 %	1.27 %
Vertical alignments of brackets	Unskewed	992	86.39 %	0.01 %	13.6 %
	Skewed	268	92.91 %	0.04 %	7.05 %
	Total	1 260	87.78 %	0.02 %	12.2 %
Total		21 103	90.61 %	0.01 %	9.38 %

Table 1. Results of line detection within French tax form.

Line type	Total number	Well classified lines	Missed lines	Bad classified lines		
				Continuous	Dashed	Dotted
Continuous	4 888	98.53 %	0.01 %	/////	1.35 %	0.01 %
Double	1 558	97.82 %	1.39 %	/////	0.78 %	0.01 %
Dashed	3 682	93.56 %	2.12 %	4.3 %	/////	0.01 %
Dotted	1 385	88.45 %	0.42 %	0.01 %	11.12 %	/////
Total	11 513	95.63 %	1.91 %	0.71 %	1.71 %	0.01 %

Table 2. Results of line detection within tables.

Items	Total	Well found items	Under-segmented items	Over-segmented items	Forgotten items
French Tax Forms	5 147	94.7 %	4.51 %	0.05 %	0.74 %
Tables	6 901	85.57 %	9.14 %	0.88 %	4.41 %

Table 3. Results of item extraction.

Fig. 8 shows the found items in a French tax form page. The results of item extraction are described in Table 3. Under-segmentation comes from item delimiters that are too small to be detected. The forgotten items often have rounded vertices. Characters which are lined up with a Hough line and so close to two other lines than they are considered wrongly as a continuous line, cause over-segmentation (see the upper item in Fig. 8). All these errors will be corrected by specific treatment.

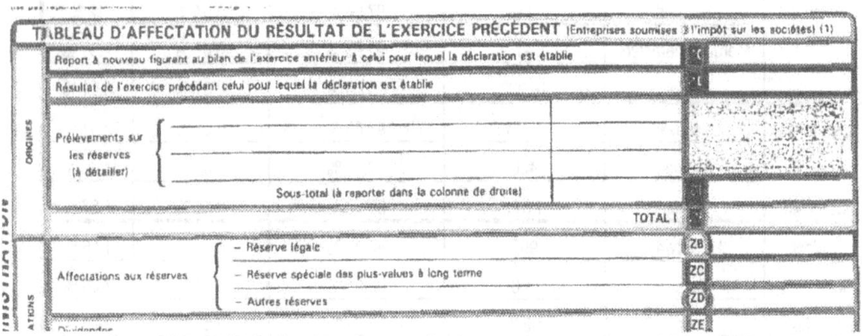

Fig. 8. Found items in an extract of a French tax form.

5 Conclusion

This paper propose a framework of form and table analysis without using any *a priori* knowledge on the item location. The idea is to have a general tool adaptable for different forms and from which results can be easily integrated into different systems designed for specific interpretations. So, an item searching method based on crossing line analysis has been developed. Lines are detected by Hough transform and segments are located as item sides. The item searching is founded on the hypothesis that an item is surrounded by a polygonal line. So, a graph from crossing lines is constructed and items searched as the minimum cycles of this graph. According to the same hypothesis made before, lines are finally classified with a homogeneous type: continuous, dashed or dotted.

The results on French tax forms and tables demonstrate that this system is robust in regard to various types of lines and reliable enough to be integrated in more specific systems to further investigations. In fact, the low rate of errors coming from: small lines, connected characters, bad classification of dotted lines and rounded vertices, can be easily omitted. In the first case, additional knowledge of line dimensions will be needed. The other cases will be solved by the examination of line context, for example, the line shape (horizontal or vertical extension of the line due to the presence of a merged character). For round vertices, the knowledge of their specific location at graph corners allowsthe use of an adaptable following process.

References

1. Yuan, J., Xu, L.,and Suen, C.Y.: Form Items Extraction by Model matching. *First International Conference on Document Analysis and Recognition*, vol. 1, pages 210-218, October 1991.

2. Yan, C.D., Tang, Y.Y.and Suen, C.Y.: Form Understanding System Based on Form Description Language. *First International Conference on Document Analysis and Recognition*, vol. 1, pages 283-293, October 1991.

3. Wong, K.Y., Casey, R.G., and Wahl, F.M.: Document Analysis System, *IBM Journal. Research and Development.*, vol. 26, November 1982.

4. Watanabe, T., Naruse, H., Luo Q., and Sugie, N.: Structure Analysis of Table-form Documents on the Basis of the Recognition of Vertical and Horizontal Line Segments. *First International Conference on Document Analysis and Recognition*, vol. 2, pages 638-646, October 1991.

5. Taylor, S., Fritzson, R., and Pastor, J.A.: Extraction of Data from Preprinted Form. *Machine Vision and Applications*, vol. 5, pages 211-222, 1992,

6. Lam, S.W., Javanbakht, L., and Srihari, S.N.: Anatomy of a Form Reader. *Second International Conference on Document Analysis and Recognition*, pages 506-509, October 1993.

7. Illingworth, J., and Kittler, J.: A survey of the Hough transform. *Computer Vision, Graphics, and Image Processing*, vol. 44, pages 87-117, 1988.

8. Risse, T.: Hough Transform for Line Recognition : Complexity of Evidence Accumulation and Cluster Detection. *Computer Vision, Graphics, and Image Processing*, vol. 46, pages 327-345, 1989.

9. Muller-Belaïd, Y., and Mohr, R.: Planes and quadrics detection using Hough transform. *7th International Conference on Pattern Recognition*; August 1984.

Model-Based Analysis of Printed Tables

Edward A. Green* and Mukki S. Krishnamoorthy

Department of Computer Science
Rensselaer Polytechnic Institute, Troy, NY 12180
greene@cs.rpi.edu

Abstract. We discuss our system of model-based analysis of printed tables. The goal of our system is to extract and associate parts of a table's image into related segments. For example, we can locate columns, rows, column and table headings of a table's image. The location of these segments are based on features of the table image and on a model of the table. This is a stepwise, top-down approach to table image analysis; thus, for example, the body of the table is located before rows or columns, or individual table cells. The algorithms we discuss involve the stepwise analysis of the image and the grouping of these segments into larger structures (columns, rows, etc.).

Keywords: tables, model-based analysis, top-down analysis.

1 Introduction

Tables may be printed in many different styles. Aside from the various devices used to distinguish one cell from another (white space, ruling lines of varying widths, for example) a table may have different orientations. For example, tables may be organized such that each row in the body represents one item, with the items' attributes specified in the columns. Another typical arrangement transposes this table, so that items are arranged in columns. A third common organization has each cell (rather than rows or columns of cells) representing an item (for example, the Periodic Chart of the Elements) [1]. At the most abstract level, of course, a table is a relation. A system that attempts to understand a table is actually attempting to derive the relational information stored in the printed manifestation of the table.

Most previous systems for analyzing tables have concentrated on analyzing tables with some specific format or restricted set of formats [1][2][3]. In this paper we describe a system that can analyze a wide variety of printed table formats. The adaptability of this system is realized by a model of the table's organization.

Printed representations of relational data rely on several kinds of visual clues for imparting the table's logical structure to the reader. For example, ruling lines of various widths might indicate a grouping of consecutive items or attributes. A system reading a table must make deductions based on these visual devices before it is able to specify the relational organization of the table. Some of the visual clues used to logically organize the physically structured information in the table are:

. *ruling line locations and attributes (e.g., thickness)*
. *font characteristics of cell contents (e.g., bold-face type)*
. *the semantics of the cell text*
. *format characteristics of cell data (e.g., integer vs. real)*
. *the shape of the cells (i.e., the cells aspect ratio)*

The approach asserted by this research is that the table analysis task can be done in steps, from analyzing the table's image to the logical interpretation of the table. There are two phases. The first phase generates a set of markers from the table image which would serve to logically organize the space occupied by the table; in other words, the first phase segments the table image. This segmentation would be expressed in a manner to allow efficient extraction of the logical relationships represented by the table. The second phase takes the segmented table and extracts the logically related data.

This paper discusses the algorithms used in these two phases. Section 2 of this paper diagrams the overall system design. Section 3 describes cell labels. Cell labels are used in each cycle of the analysis and for matching cells with templates to extract logic. An algorithm is provided which generates a labeling of individual cells. Section 4 describes how logically related cells may be grouped using cell labels. Section 5 describes the components of a table model. Section 6 describes some experimental results and Section 7 concludes the paper.

2 System Design

The overall design of our system is diagramed in figure 1:

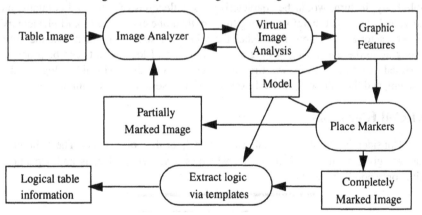

Fig. 1: Table analysis system

The features of the table image are found by the image analyzer. The virtual image analysis provides a "front end" to the image analyzer: it translates commands to find features in specific areas and translates the results into a form that the rest of the system can use. Since the actual image analyzer retains independence from the rest of the system, new feature extraction programs can be substituted without affecting the rest of the system. The features are simply characterized by their bounding rectangle and a label: all separator structures will have rectangular shapes if the image is (as we are assuming) initially deskewed.

Currently, only features based on white space and solid ruling lines are found and used. This is sufficient for analyzing many types of tables. Attributes include the width of white space and whether ruling lines are single or double, etc. The image analysis is based on connected components and is susceptible to noise and skew.

The image analysis and virtual image analysis locate the graphic features which form the basis of segmenting or "marking" the table. Marking the table is a cyclic process; first the table is found on the page, then the table heading, column headings, and table body are segmented (for example), and finally individual cells are recognized. Each cycle requires that features be found on which to base the segmentation. The features which will be used at each cycle of the process and in each location of the table are specified in the model (the topic of section 5).

The final marked image is then examined for logical features (also specified in the model). Thus, the area of the table corresponding to (for example) "tuple 1" or "column B" can be recognized. Templates are provided by the model for this purpose. Thus, the model coordinates feature extraction and logical assignment of the resulting segments.

3 Characterizing Table Cells

The characterization of physical cells is hierarchical: at the top level, all table cells exist in the table. Individual cells belong to distinct (but possibly overlapping) logical classes: table header, column headings, and table body, for instance. The table body, in turn, would be characterized as collections of rows and columns; an individual cell might therefore belong to more than one class. The task of elementary cell characterization, then, is to label these cells in some way such that the cells belonging to the underlying nesting and overlapping of logical units can be properly extracted. In this section we first consider the one-dimensional analog of the problem, and then show a practical extension of this solution to two dimensions.

3.1 Cell Hierarchies

Consider a one dimensional, say vertical, profile of a table. The following diagram gives such a vertical profile, which is a sequence of cells and separators with corresponding weights.

Fig. 2: A one-dimensional cell structure

This sequence of cells (represented by c) and separators might be represented by the string:

$$4c2c3c2c3c1c1c4$$

or, if []=4, {}=3, ()=2, and ε =1, then the string is represented in a more familiar notation:

$$[c)(c]\{c)(c\}\{ccc]$$

For a "proper" nesting, the brackets, braces, and parenthesis should be balanced, and each cell should be at the bottom-most level (within parenthesis), so the string should be converted to:

$$[\{(c)(c)\}\{(c)(c)\}\{(ccc)\}]$$

This is a straight-forward transformation in one dimension; since parenthesis, braces, and brackets are not nested in this formulation, it is a regular transformation. A regular grammar which generates this "balanced cell language" is:

$$S \rightarrow [A$$
$$A \rightarrow \{B$$
$$B \rightarrow (C$$
$$C \rightarrow cD$$
$$D \rightarrow cD \mid)E \mid)B$$
$$E \rightarrow \}F \mid \}A$$
$$F \rightarrow]$$

The extension to two dimensions is straightforward. for example, consider figure 3.

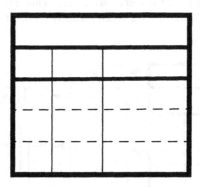

Fig. 3: A two-dimensional cell structure (i.e., a table).

The natural approach for handling the two-dimensional case is to consider regions of cells in the two- dimensional grid, and to group those cells separated by the lower weight separators within the compound cells surrounded by heavier weight separators. Note that this is identical to nesting each cell surrounded by light-weight separators inside cells surrounded by heavier weight separators, resulting in a contour map of nested cell hierarchies.

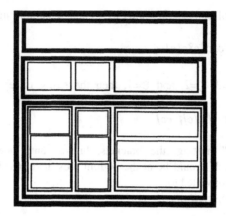

Fig. 4: Contour map of cell hierarchies.

Given this natural hierarchical structuring of a table, we now address the problem of referencing individual cells in the table in a way that reflects the cell's position in the hierarchy.

3.2 Cell labeling format

The notation used for our cell labeling is an extension of the familiar notation for cell addresses used in many popular spreadsheets. Cells are addressed based on its column and row location. The upper left ("home") address is A1. Columns run from addresses A through Z, then AA through AZ, then BA through BZ, etc. The rows are numbered from 1. The two cells which are "knight moves" from the home address are, therefore, C2 and B3. C2 is a diagonally adjacent cell to (up and to the right of) cell B3.

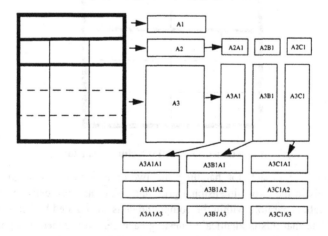

Fig. 5: Labeling a table

The labeling system developed here extends the normal spreadsheet labeling by concatenating spreadsheet cell labels, one label for each level in the hierarchy: for example, a typical address of an individual cell in a five level hierarchy would be A1C1A3B1A1. A cell group at some level in the hierarchy is addressed relative to its position in the enclosing next higher level. Since groups of cells exist in each level of the hierarchy (except the bottom level), a group of cells are referenced relative to the position of the top left-most cell in a group (see figure 5).

3.3 Algorithm to Label Cells

Cells are grouped by building a general tree with the assistance of a minimum priority queue.

1. Start with any individual cell in the table. This is the first "current" cell.

2. Initialize a tree to be n nodes in height, with a single leaf, where n is the number of separator weights. Attach the current cell to the leaf of this tree.

3. Mark the current cell (so it will no be selected again).

4. Place each unmarked neighbor of the current cell into a priority queue. The key is the separator weight between the current cell and the neighbor.

5. Pull the minimum weight cell from the queue. Let this be the "new" cell. (If the queue is empty, exit.)

6. If the new cell is marked, goto 5.

7. The current cell belongs to n levels. Let k be the key of the new cell. Perform the following operations on the tree:
    > ascend the tree n-k levels.
    > add a new subtree to this level. This subtree will be a single leaf tree of n-k height.
    > attach the new cell to the new leaf.

8. The new cell becomes the current cell.

9. Goto 3.

Once the tree is built, (recursively) sort the child nodes of each level according to the upper-left most position of the levels beneath. Each level of the tree then represents one part of the label of the child nodes beneath it, and labels can be generated by traversing the tree.

4 Templates

At this point, the labeling of individual cells relative to a separator structure has been achieved. The goal is to match this raw data to a model such that relational information in the table can be extracted.

A template is a set of logical specifications or assignments. These assignments map a logical label to a set of cells in the table: all logical entities in a table will be represented by some set of individual cells. A logical specification in this project has the following format:

$$<logical\ label> = <cell\ label\ template>$$

The explanation of templates is easier with reference to an actual example.

A1		
A2A1	A2B1	A2C1
A3A1A1	A3B1A1	A3C1A1
A3A1A2	A3B1A2	A3C1A2
A3A1A3	A3B1A3	A3C1A2

Fig. 6: An example labeled table.

There are 3 levels of separator hierarchy in this example. If the first row of is considered to be a table heading, then the logical assignment would be:

table heading = A1

Wild-carding allows multiple cells to be assigned to a single logical entity:

column headings = A2??

Where "?" is the wild card heading. This matches cells A2A1, A2B1, and A2C1, the column headings.

A simple extension allows a multiple assignment, based on a wild card reference in the logical label:

col?2 = A3??

groups all elementary cells matching the template by the 2nd coordinate("?2"), numbering from the 0th cell label template position (the "A"). This specification would be the same as the three specifications:

colA = A3A????
colB = A3B????
colC = A3C????

However, if these later specifications were specified in a model file, the model would only be able to characterize 3 column tables. Using multiple assignments allows for a varying number of columns to be handled by a model.

5 The Table Model

The core of our system is the table model. The table model supplies the essential characteristics that a table will exhibit if the table can be recognized, and coordinates the mapping of graphical attributes to logical structure. The model is divided into 4 main parts: feature codes, image isolation, phases, and templates.

This section describes each section, and shows an example of analyzing a table using the model. The example table is shown in figure 7.

Table 1: Noise Filtering Results Total Dots = 171 (manual count) Total Noise specks = 73		
Filter Size	Dots Erased	Noise Erased.
0x0	0	0
1x1	0	30
2x2	34	40
3x3	154	55
4x4	171	67
5x5	171	73

Fig. 7: Example table to illustrate table models.

5.1 Feature codes

Feature codes indicate all of the graphical features that are required to separate each cell. A few examples of features and feature codes are horizontal lines (HL), vertical lines (VL), horizontal space (HS), and vertical space (VS). Feature codes must be recognized by the virtual image analyzer. Other codes may be added as required. The entry in the model file for this table is:

[feature-codes]

    HL VL HS HL-M

where HL-M means that analyzing the table requires recognition of multiple horizontal lines.

5.2 Image isolation

These are the features of a table which separate the table from the remainder of the image. In this work, we assume that the table image lies on a white background. Thus, white space features (HS and VS) can be used. If the table is surrounded by lines (as in a box), then horizontal and vertical lines can be used. As shown in section 3, this will form the "heaviest weight" separator.

For the example table (which is in a box), the entry for image isolation would be:

[image-isolation]

    * top bottom left right (comments start with *)

    HL-M HL VL VL

5.3 Phases

As we proceed in a top down fashion, each phase of marker placement subdivides the areas marked in the previous phase. For a given phase, each cell of the previous phase may be subdivided, and different image features may used to subdivide different cells. All separators placed in a given phase will have the same weight, and each separator (which is to say, the underlying graphical feature) must span the entire cell to cause a separator to be placed. Between phases cell labels are calculated anew, and are used to specify the cells to be subdivided.

For the example table, there are 3 phases used to separate the table. The entry is:

> *[phase 0]*
>> *?: HL   * ? indicates the entire table image.*
> *[phase 1]*
>> *A2: VL*
>> *A3: VL*
> *[phase 2]*
>> *A2??: HS*

This decomposes the table in accordance with figures 5 and 6.

5.4 Templates

Templates were discussed in section 4. A set of templates which can be used to characterize the table in figure 7 might be:

> *[templates]*
> *tablehead = A1*
> *colhead?2 = A2??*
> *col?2 = A3????*
> *tuple?5 = A3????*
> ** mark end of model file*
> *[end]*

6 Results

A table analysis system with an X Window System graphical interface and based on the notions described above was written. We present in this section preliminary results of using this system on a set of tables; the system was used to analyze the table of contents of the SIAM Journal on Computing; the model was developed manually using one table (the April 1994 issue) and the model was tested on 9 other tables. A sample of the table image is shown in figure 8, and the model file is shown in figure 9.

The separators are all based on white space between entries: for example, the space between title and author is narrower than the spaces between articles; this is one basis for drawing markers. Since the feature extraction only looks at white space and ruling lines, errors will occur where there is more than one line of title or authors; extraneous marks will be drawn in this case. If feature extraction sensitive to typeface were substituted for the one used in this implementation such an error could be avoided.

With the graphical user interface used by the system these extra marks can be deleted (or missing marks inserted) during the analysis process and an accurate result can be obtained. The number of such corrections are tabulated in terms of mouse clicks needed to make the corrections: this is presented in table 1.

CONTENTS

227 A Linear-Time Algorithm for the Homotopic Routing Problem in Grid Graphs
Michael Kaufmann and Kurt Mehlhorn

247 Some Results on Elusive Graph Properties
Eberhard Triesch

255 Closeness of NP-Hard Sets to Other Complexity Classes
Bin Fu and Hong-Zhou Li

261 Complexity-Restricted Advice Functions
Johannes Köbler and Thomas Thierauf

276 Deterministic Simulations of PRAMs on Bounded Degree Networks
Kieran T. Herley and Gianfranco Bilardi

293 Lower Bounds for Randomized k-Server and Motion-Planning Algorithms
Howard Karloff, Yuval Rabani, and Yiftach Ravid

313 An Alphabet Independent Approach to Two-Dimensional Pattern Matching
Amihood Amir, Gary Benson, and Martin Farach

324 Trading Space for Time in Undirected s-t Connectivity
Andrei Z. Broder, Anna R. Karlin, Prabhakar Raghavan, and Eli Upfal

335 Bounds on the Costs of Multivalued Register Implementations
Soma Chaudhuri and Jennifer L. Welch

355 Parallel Information Dissemination by Packets
A. Bagchi, E. F. Schmeichel, and S. L. Hakimi

373 Generating Linear Extensions Fast
Gara Pruesse and Frank Ruskey

387 Randomized Parallel Algorithms for Matroid Union and Intersection, With Applications to Arborescences and Edge-Disjoint Spanning Trees
H. Narayanan, Huzur Saran, and Vijay V. Vazirani

398 On the Consecutive-Retrieval Problem
R. Swaminathan and Donald K. Wagner

415 Polynomial Root-Finding Algorithms and Branched Covers
Myong-Hi Kim and Scott Sutherland

437 Computing and Verifying Depth Orders
Mark de Berg, Mark Overmars, and Otfried Schwarzkopf

447 Erratum: Optimal Parallel Randomized Algorithms for Three-Dimensional Convex Hulls and Related Problems
John H. Reif and Sandeep Sen

Fig. 8: SIAM Journal on Computing table of contents

```
[feature-codes]
     HS VS HS-W.5 HS-W.99 VS-W.5 HS-W.0
[image-isolation]
     HS HS VS VS
[phase 0]
     ?: HS-W.90
[phase 1]
     A2:HS-W.5
[phase 2]
     A2:VS-W.5
[phase 3]
     A2??B?:HS-W.0
[templates]
Article_?3 = A2A?
Title_?3 = A2A?B1A1
Author_?3 = A2A?B1A2
Page_Num_?3 = A2A?A1
[end]
```

Fig. 9: Model for SIAM Journal on Computing table of contents

Table 1: Results of table analysis of SIAM table of contents

Month of Issue	Clicks to change edit mode	Clicks on table image
Apr 1994*	1	2
Apr 1995	1	1
Aug 1993	1	2
Dec 1993	2	3
Dec 1994	4	6
Feb 1994	1	4
Feb 1995	1	1
Jun 1995	1	4
Oct 1993	3	3
Oct 1994***	5	9
Apr 1994**	15	99

*=table used to develop model
**=no model used to analyze table - analysis done entirely by hand on the image
***=poor image quality

7 Discussion and Conclusion

The data reflected in a table image may be the result of the merging or joining of several different relations or SQL tables. The ultimate goal of this work is the reconstruction of constituent relations from table images.

Deciding on sources of table input appropriate for this system is an interesting question. Tables which are static would not be of practical importance for this system; the Periodic Table is sufficiently static such that the information reflected in the table image would already be available in an existing database; it's not important that the Periodic Table be read and analyzed every time that it is encountered. Additionally, tables which have a very fixed format and which already have a system built explicitly for them are also not appropriate for this system; for example, tables for income tax forms (at least for a given year) do not vary at all in format, so recognition systems for these tables could be hard coded; these tables don't need the power of a modeling system.

Now that we have suggested some inappropriate tables, what would constitute an appropriate source of tables? Those tables which vary somewhat in format, that is, in number of columns, various column heading formats, etc. would be good candidates for this system. Currently, we are considering tables that appear in specific scientific journals or in textbooks, since these tables usually follow a standard set of layout rules, which would make models easier to develop. The question is still open, however.

Tables within a publication may follow one of a set number of layout patterns, instead of a single pattern. The plausibility of detecting one of several general table layouts will be studied.

Simplifying table model development is a key issue in this system. Two tasks related areas of improvement are an interface for the naive user, and fine-tuning old models or deducing new models by user corrections to the table analysis (using supervised learning techniques).

References

[1] Laurentini, A. and P. Viada, "Identifying and Understanding Tabular Material in Compound Documents," Proc. ICPR, 1992, pp. 405-409.

[2] Chandran, S. and R. Kasturi, "Structural Recognition of Tabulated Data," ICDAR Japan, October 1993, pp. 516-519.

[3] Watanabe, T., H. Naruse, Q. Luo, N. Sugie, "Structure Analysis of Table-form Documents on the Basis of the Recognition of Vertical and Horizontal Line Segments," ICDAR France, 1991, pp. 638-646.

Morphological Approach for Dashed Lines Detection*

Gady Agam, Huizhu Luo and Its'hak Dinstein

Department of Electrical and Computer Engineering, Ben-Gurion University of the Negev, Beer-Sheva 84105, Israel

Abstract. New directional morphological operators that have accurate selectivity and controllable strictness, are defined and applied to dashed lines detection and labeling. The proposed approach is based on adaptation of the directions and dimensions of newly defined tube-directional morphological operators to local characteristics of the data. The separation of maps and line drawings, into four images containing respectively graphics, character strings, symbols, and dashes was presented in a previous paper. This paper presents an algorithm for detection and labeling of dashed lines, where the input is an image of dashes. Experimental results demonstrate very good detection even in cases where the dashed lines intersect.

1   Introduction

The problem of dashed lines detection is a common problem in many computerized document analysis applications [3], such as the analysis of engineering drawings and maps. Given a line drawing image, it is possible to distinguish in it four basic types of objects: lines, dashed lines, character strings, and symbols. In the proposed approach, lines are separated from all the other objects, then character strings are restored and separated, and finally symbols are separated from dashes. The dashes image is then analyzed in order to detect and label dashed lines in it. This paper discusses the problem of dashed lines detection in the dashes image, where the method for removal of lines, character strings, and symbols is presented in another paper [4]. The dashes of dashed lines are considered as segment-samples of discrete full lines, and so in order to recognize dashed lines it is required to reconstruct the respective full lines. After the reconstruction of the dashed lines, it is possible to recognize them by applying the same means that are used for full lines recognition. For the purpose of dashed lines reconstruction, new directional morphological operators are defined, based on concepts of: directional selectivity, controllable strictness, and local adaptivity.

The problem of dashed lines detection is handled in many document analysis systems [5, 2], by treating the dashes of dashed lines as a collection of separate

* This work was partially supported by The Paul Ivanier Center for Robotics and Production Automation, Ben-Gurion University of the Negev, Beer-Sheva 84105, Israel.

objects of short lines. Such representation however, is less accurate and less compact than a representation that combines many dashes into a single dashed line. In addition, by detecting a dashed line object, it is possible to identify and use its functionality. The system described in [3], performs dashed lines detection by checking all the possible connections between each dash to all the other dashes, and then taking decision based on an average of near neighbors. In this method, the size of the search space increases with the number of dashes in the image, and errors may occur when the average is diverted by close dashed lines. In addition, this method impose strong constraints on the possible structure of the dashed lines.

The following sections discuss the dashed lines reconstruction problem and the proposed approach. Description of directional edge planes is presented in Section 2. Directional morphological operators are described in Section 3, where Subsection 3.1 presents simple directional operators and Subsection 3.2 defines tube-directional operators. Section 4 defines adaptive directional morphological operators, where Subsection 4.1 presents the method for extraction of adaptation characteristics, and Subsection 4.2 describes the method in which the tube parameters are adapted. Finally Section 5 presents the dashed lines reconstruction process. The summary in Section 6 concludes the paper.

2   Directional Edge Planes

Given a binary image, objects in it (like lines, characters, or symbols) may be represented more efficiently by their contours, that is by the sets of their edge pixels. Since the number of pixels in the contour of an object is smaller then the number of pixels in the object itself, the amount of data that has to be processed when using a contour representation, is smaller. In order to enable efficient processing with directional morphology operators, the original contours are decomposed into eight directional edge planes. After the decomposition, the direction of the obtained edges in the same directional edge plane is fixed, and so the same directional morphological kernel may be applied for all the edges in that plane, without needing to determine the direction of each edge.

Decomposition into directional edge planes [8, 4] has the feature of stressing directional information in the image such as direction of lines and strings, and so gets support from the human visual system. Therefore by using directional edge planes, ambiguities in the image are likely to be resolved in a more natural way. An example to ambiguity resolved by the use of directional edge planes is presented in Fig. 1. Figure 1-a presents an image with two intersecting dashed lines where since dash 1 is close to dashes 2-4, there is ambiguity concerning to which of these dashes should dash 1 be connected to. Figure 1-b presents a directional plane in which there is no ambiguity since dash 1 can be only connected to dash 3. That is the ambiguity of the junction in this example is resolved due to collinearity introduced by the directional plane.

In the following description 8-neighborhood of a pixel is assumed, where the right neighbor is in direction 0, and the rest of the directions are numbered

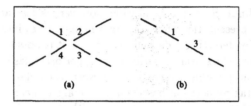

Fig. 1. Example of ambiguity resolved in a directional plane

sequentially clockwise. It is also assumed that the direction indexes are always modulo-8. Given a binary image $F \equiv \{f(i,j)\}$, the directional edge planes in the vertical and horizontal directions may be obtained by:

$$F_d^c \equiv F \cap \overline{F[d+4]} \ , \quad d = 0, 2, 4, 6 \tag{1}$$

where $F[d] \equiv \{f[d](i,j)\}$ denotes the image F shifted by one pixel towards direction $(d+4)$ (so that $f[d](i,j)$ is the d-th neighbor of $f(i,j)$), and the overline denotes inverse (one's complement). After applying this operation the horizontal and vertical directional edge planes may contain diagonal edges. These diagonal edges can be separated from the horizontal and vertical edge planes and then combined to create the diagonal directional edge planes, by issuing:

$$F_d^e \equiv F_d^c \cap (F_d^c[d-2] \cup F_d^c[d+2]) \tag{2}$$

for $d = 0, 2, 4, 6$, and

$$F_d^e \equiv F_{d-1}^c \cap (F_{d-1}^c[d-2] \cup F_{d-1}^c[d+2]) \cup F_{d+1}^c \cap (F_{d+1}^c[d-2] \cup F_{d+1}^c[d+2]) \tag{3}$$

for $d = 1, 3, 5, 7$. Figure 2 presents an example for directional edge planes extraction, where Fig. 2-a presents the original image, and Figs. 2-b – 2-e present the extracted directional edge planes in directions 0–3 respectively. It should be noted that according to the definition of the directional edge planes, the directional edge plane F_d^e contains edge segments in direction perpendicular to direction d (that is in direction $d + 2$).

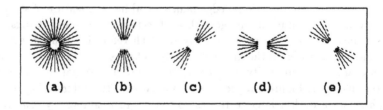

Fig. 2. Extraction of directional edge planes

An additional property of decomposition into directional edge planes is that curved edge lines are decomposed into several edge segments in different directional edge planes, where each edge segment actually forms a side in some kind of a polygonal approximation [6] of the curved edge. This property may be later used for the vectorization of lines in the image.

3 Directional Morphology

3.1 Simple Directional Operators

Directional morphological operators [8, 4] may be classified as a subset of general morphological operators, in which the morphological kernels that are used are non-isotropic and so give preference to some direction. The simplest directional morphological operator, may be obtained by using a kernel consisting of the origin pixel and one 8-connected neighbor. Such morphological operator is called single-directional operator. The single directional dilation and erosion may be expressed by: $D_{=d}F \equiv F \cup F[d+4]$ and $E_{=d}F \equiv F \cap F[d]$ respectively, where d determines the direction of the operators. In these definitions the shift of F in the dilation operator is opposite to its shift in the erosion operator, so that line dilated from one of its edges by $D_{=d}$ is eroded by $E_{=d}$ from the same edge.

Since the propagation in one application of a single-directional morphological operator is small, these operators are usually applied successively. A successive application of a morphological operator is denoted here by a superscript on the morphological operator, and so for example $D^n_{=d}$ denotes n successive applications of $D_{=d}$. The propagation path of a single-directional operator, created by successive application of it, is presented in Fig. 3-b where for comparison Fig. 3-a presents the propagation path of simple non-directional operators defined by: $DF \equiv \bigcup_{d=0}^{7} D_{=d}F$ and $EF \equiv \bigcap_{d=0}^{7} E_{=d}F$.

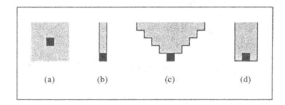

(a)　　　　(b)　　　　(c)　　　　(d)

Fig. 3. Propagation path of morphological operators

The two major problems involved with single-directional morphological operators are that they are too selective and so they are sensitive to noise and artifacts of digital lines, and that their possible direction is limited to a small set of discrete angles (multiples of 45°). In order to solve the selectivity problem of single-directional morphological operators, fan-directional morphological operators are defined by using a kernel consisting of the origin pixel and three consecutive 8-connected neighbors in the required direction. Fan-directional morphological operators may be expressed by:

$$D_{>d}F \equiv F \cup \left(\bigcup_{\delta=d+3}^{d+5} F[\delta] \right) \quad ; \quad E_{>d}F \equiv F \cap \left(\bigcup_{\delta=d-1}^{d+1} F[\delta] \right) \tag{4}$$

It should be noted that here, like in the definitions of single-directional operators, the shift in the erosion operator is opposite to the shift in the dilation operator, from the same reasons. It should also be noted that in the definition of $E_{>d}$ the union operator is used, and so in order for a pixel to be left in the eroded image, it has to have at least one neighbor in the respective kernel area. A definition of erosion with an intersection operator would result in an excessive erosion, since it would require to have all the neighbors present in the kernel area. The propagation path of fan-directional operators is presented in Fig. 3-c.

3.2 Tube Directional Operators

The major drawback of fan-directional operators is that their selectivity decreases with the increase of the number of their successive application (since their propagation path is a cone). In addition, the problem of the limited number of discrete basic angles still exist. The tube-directional morphological operators are designed to overcome these problems. A tube-directional morphological operator is basically obtained by moving a simple bar kernel along a discrete propagation line in a required angle, and so selectivity is determined exactly by the tube width, and the direction of propagation is not limited to a small set of discrete angles.

Given a binary image $F \equiv \{f(i,j)\}$ with m rows and n columns, a tube-directional morphological dilation operator in direction θ with width W, length L, and strictness S, is defined by:

$$D_{\theta,W,S}^L f(i,j) \equiv \begin{cases} 1 \text{ if } f(i,j) = 1 \ \vee \\ \sum_{\substack{(k,l) \in T(\theta+180,W,L): \\ (i+k,j+l) \in \mathcal{I}_F}} f(i+k,j+l) \ \geq \ \tau_D(S,\theta,W,L) \\ 0 \text{ otherwise} \end{cases}$$
$$\tag{5}$$

where the threshold τ_D is given by:

$$\tau_D(S,\theta,W,L) \equiv S \cdot (\#T(\theta+180,W,L) - 1) \tag{6}$$

the symbol $\#$ denotes cardinality of a set, $T(\theta+180,W,L)$ is the set of the required tube kernel pixels, and $\mathcal{I}_F$ is the set of all the pixels in the image given by the Cartesian product:

$$\mathcal{I}_F \equiv \{0,\ldots,m-1\} \times \{0,\ldots,n-1\} \tag{7}$$

The set of tube kernel pixels $T(\theta, W, L)$ is defined by:

$$T(\theta, W, L) \equiv \{(i, j + k) \mid (i, j) \in \mathcal{L}_{0, P_e(0, \theta, L)} \land k \in [-H, H]\} \qquad (8)$$

if $\theta \in [45, 135] \lor \theta \in [225, 315]$, and by:

$$T(\theta, W, L) \equiv \{(i + k, j) \mid (i, j) \in \mathcal{L}_{0, P_e(0, \theta, L)} \land k \in [-H, H]\} \qquad (9)$$

otherwise, where $H \equiv \lfloor \frac{W}{2} \rfloor$, $\mathcal{L}_{0, P_e(0, \theta, L)}$ is a set of pixels of the required discrete propagation line, and $P_e(0, \theta, L)$ is the ending pixel of that line. The discrete line $\mathcal{L}_{P_s, P_e}$ that connects between the pixels $P_s \equiv (i_s, j_s)$ and $P_e \equiv (i_e, j_e)$, assuming that $P_s \neq P_e$, is given by:

$$\mathcal{L}_{P_s, P_e} \equiv \{(i, j) \in \mathcal{I}_F \mid j_{min} \leq j \leq j_{max} , \ i = \lfloor \frac{d_i}{d_j}(j - j_s) + i_s + \frac{1}{2} \rfloor\} \qquad (10)$$

when $\mid d_j \mid \geq \mid d_i \mid$, and by:

$$\mathcal{L}_{P_s, P_e} \equiv \{(i, j) \in \mathcal{I}_F \mid i_{min} \leq i \leq i_{max} , \ j = \lfloor \frac{d_j}{d_i}(i - i_s) + j_s + \frac{1}{2} \rfloor\} \qquad (11)$$

otherwise, where $i_{min} \equiv \min(i_s, i_e)$, $i_{max} \equiv \max(i_s, i_e)$, $d_i \equiv i_e - i_s$, and j_{min}, j_{max}, d_j are defined similarly by using j_s and j_e. The ending pixel $P_e(P_s, \theta, L)$ of a discrete line starting at P_s with angle θ and length L is given by:

$$P_e(P_s, \theta, L) \equiv P_s + (-L\sin(\theta), L\cos(\theta)) \qquad (12)$$

The definition of tube-directional erosion is done similarly to the definition of tube-directional dilation, where instead of requiring the sum to be greater than a required threshold in order to turn a pixel to 1, the condition in the definition of tube-directional erosion requires that the sum should be smaller than a required threshold in order to turn a pixel to 0. A tube-directional morphological erosion operator in direction θ with width W, length L, and strictness S, is defined by:

$$E^L_{\theta, W, S} f(i, j) \equiv \begin{cases} 0 \text{ if } f(i, j) = 0 \ \lor \\ \\ \sum_{\substack{(k, l) \in T(\theta, W, L) : \\ (i + k, j + l) \in \mathcal{I}_F}} f(i + k, j + l) \ \leq \ \tau_E(S, \theta, W, L) \\ \\ 1 \text{ otherwise} \end{cases}$$

$$(13)$$

where the threshold τ_E is given by:

$$\tau_E(S, \theta, W, L) \equiv S \cdot (\#T(\theta, W, L) - 1) \qquad (14)$$

It should be noted that the length parameter L is actually equivalent to the number of successive applications of simple directional operators, since it determines the number of moves of the simple bar kernel along the propagation line. It should also be noted that here, like in the definitions of single-directional operators, the propagation in the erosion operator is opposite to the propagation in the dilation operator, from the same reasons. The propagation path of tube-directional operators is presented in Fig. 3-d.

The strictness parameter S in the definitions of the tube-directional morphological operators control the number of neighbors that have to be present in the kernel area, where for D^L_{θ,W,S_D} and E^L_{θ,W,S_E} these parameters are limited to:

$$S_D \in \left[\frac{1}{\#\mathcal{T}(\theta+180,W,L)-1} , 1 \right] \quad ; \quad S_E \in \left[\frac{1}{\#\mathcal{T}(\theta,W,L)-1} , 1 \right] \quad (15)$$

When S_D or S_E are maximal the strictness is maximal since all the neighbors should be present in the kernel area, and when S_D or S_E are minimal the strictness is minimal since only one neighbor has to be present in the kernel area. The strictness parameter is introduced in order to prevent excessive dilation or erosion when using large kernels, by treating the kernel area as a search area in which a certain number of neighbors have to be present. Excessive dilation is prevented by increasing the required strictness S_D, and excessive erosion is prevented by decreasing the required strictness S_E. It should be noted that the parameter of strictness in the tube erosion operator actually compromises between the very strict attitude taken by the non-directional erosion operator EF, and the very loose attitude taken by the fan erosion operator $E_{>d}F$.

The basic tube-directional morphological operators may be combined to create new operators such as tube-directional open given by:

$$O^L_{\theta,W,S_D,S_E}F \equiv D^L_{\theta,W,S_D} E^L_{\theta,W,S_E}F \quad (16)$$

and tube-directional close given by:

$$C^L_{\theta,W,S_E,S_D}F \equiv E^L_{\theta,W,S_E} D^L_{\theta,W,S_D}F \quad (17)$$

As could be noted, the strictness of erosion and the strictness of dilation are not necessarily the same. Figure 4 presents results obtained by applying tube operators. Figure 4-a presents the original image, Fig. 4-b presents results of tube erosion in direction of 45°, and Fig. 4-c presents results of tube open in direction of 45°.

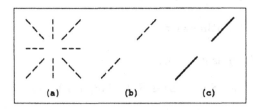

Fig. 4. Demonstration of tube-directional morphological operators

4 Adaptive Directional Morphology

In the normal sense of morphology, a kernel with a specific structure is used to traverse an image, where at each location it is intersected or unified with the image. The traversing kernel is fixed and does not depend on the contents of the traversed image. Adaptation to a specific task is achieved globally only by setting the structure of the kernel, and so for example if directional information is required the kernel structure is set to be non-isotropic. The concept of one global adjustment for many local operations is somehow conflicting, since it could be more reasonable to have a unique local adjustment for each local operation. That is in other words, to have a smart kernel that can sense the local environment and adapt itself to it accordingly, while traversing the image.

Local adaptation of the kernel structure is not necessarily possible in the general case, since the image could be such that it is not possible to determine a local kernel structure from a local neighborhood. The size and shape of the neighborhood according to which the kernel is adapted need not be similar to the structure of the kernel itself. However, the neighborhood should be small enough to keep locality and big enough to capture the structure according to which the kernel should be adapted. When using directional edge planes, the adaptation of the kernel structure may be based on the direction and size of edge line segments. That is, when tube morphological operators are used, the width, height, and direction of the tube operators may be determined locally based on the edge line segments.

4.1 Extraction of Adaptation Characteristics

In the following description it is assumed that the edge line segments in the directional edge planes are represented by r-tuples of pixels. The l-th edge line segment is denoted by:

$$C^l \equiv (P_1^l, \ldots, P_{r_l}^l) \quad , \quad P_k^l \equiv (i_k^l, j_k^l) \tag{18}$$

where r_l is the number of pixels in it. Given an edge line segment C^l, three characteristics are evaluated for it: a length characteristic $\Lambda(C^l)$, an angle characteristic $\Theta(C^l)$, and a deviation characteristic $\Delta(C^l)$. The length characteristic is defined as the length of the line connecting the edge line segment edges, and may be evaluated by: $\Lambda(C^l) \equiv \|P_{r_l}^l - P_1^l\|$. The angle characteristic is defined as the angle of the line connecting the edge line segment edges, and may be evaluated by:

$$\Theta(C^l) \equiv \begin{cases} 90 & \text{if } j_1^l = j_{r_l}^l \\ |\arctan\left(\frac{d_i^l}{d_j^l}\right)| & \text{if } sgn(d_i^l) \cdot sgn(d_j^l) < 0 \\ 180 - |\arctan\left(\frac{d_i^l}{d_j^l}\right)| & \text{otherwise} \end{cases} \tag{19}$$

where $d_i^l \equiv i_{r_l}^l - i_1^l$, $d_j^l \equiv j_{r_l}^l - j_1^l$, and $sgn(x)$ is equal to 1 when $x \geq 0$, and equal to -1 otherwise. The deviation characteristic is defined as the maximal deviation from the line connecting the edge line segment edges, and may be evaluated by:

$$\Delta(C^l) \equiv \max \left\{ d \; \middle| \; d = \left| \frac{\|(P_{r_l}^l - P_1^l) \times (P - P_1^l)\|}{\|P_{r_l}^l - P_1^l\|} \right| \; , \; P \in C^l \right\} \qquad (20)$$

where $\times$ represents vector product. Other possible deviation characteristics may be found in [1].

4.2 Adaptation of Tube Parameters

The length, angle, and deviation characteristics of an edge line segment may be used to determine the parameters θ, L, and W of a tube-directional morphological operator that has to be applied on that edge line segment. Given the edge line segment C^l and required initial tube parameters $L = L_1$ and $W = W_1$, the angle of a tube operator for that edge line segment is determined by $\theta = \Theta(C^l)$, and the length and width of that tube operator are adapted according to $\Lambda(C^l)$ and $\Delta(C^l)$. The adaptation of the tube length and width for an edge line segment is done in two stages according to the confidence in that edge line segment, where a long and narrow tube reflects good confidence, and a short and wide tube reflects poor confidence since the directionality in such a tube is fuzzified.

In the first stage of the tube dimensions adaptation, the adaptation is based on $\Lambda(C^l)$. Since long edge line segments imply more confidence, tubes for long edge line segments are set to be longer than tubes for short edge line segments. The adaptation of the tube length is achieved by: $L_2 = \mu(L_1, \Lambda(C^l))$, where μ is a non-linear mapping function that limits the maximal size of the tube length. The mapping function μ is given by:

$$\mu(y, x) \equiv y \cdot \left(1 + \exp \left(-9.2 \frac{x - x_{av}}{x_{bw}} \right) \right)^{-1} \qquad (21)$$

where x_{av} defines the transition point (the point in which $\mu = 0.5y$), and x_{bw} defines the width of the transition band (the range in which $0.01y \leq \mu \leq 0.99y$). The constants x_{av}, and x_{bw} may be obtained automatically from a length histogram of the edge line segments. The tube width parameter is then changed respectively by: $W_2 = \frac{L_2}{L_1} W_1$, so that the aspect ratio of the tube sides is retained.

In the second stage of the tube dimensions adaptation, the adaptation is based on $\Delta(C^l)$. Since large deviation in an edge line segment imply less confidence in the direction of that edge line segment, tubes for edge line segments with large deviations are set to be shorter and wider than tubes for edge line segments with small deviations. The adaptation of the tube width is achieved by: $W_3 = W_2 + k_1 \cdot \Delta(C^l)$, where k_1 is a normalization constant. The tube length parameter is then changed respectively by: $L_3 = \frac{W_2}{W_3} L_2$, so that an increase in the tube width results in a respective decrease in the tube length.

5 Dashed Lines Reconstruction

Given a binary image F of dashes, it is required to reconstruct the dashed lines in it into full lines. The process of dashed lines reconstruction, begins by decomposing the binary image F into four directional edge planes F_d^e where $d = 0, \ldots, 3$. Given a directional edge plane F_d^e, a list of $(n_d - n_{d-1})$ edge line segments denoted by $\{C^l\}_{l=(n_{d-1}+1)}^{n_d}$ is extracted for it, where n_d represents the highest index of an edge line segment in F_d^e, and $n_{(-1)} \equiv 0$. Since the direction of the edge line segments in a directional edge plane is known, the component labeling process that generates the edge line segments, may be very simple. After the extraction of edge line segments, the process of dashed lines reconstruction may be viewed as a process in which some edge line segments are connected to others.

For the purpose of deciding which of the edge line segments have to be connected, only the edges of the edge line segments are needed to be considered. Therefore a new binary image F_d^p is created, in which only the edges of the edge line segments are present. The definition of F_d^p is given by:

$$F_d^p(i, j) \equiv \begin{cases} 1 \text{ if } \exists \, l \in [n_{d-1}+1 \, , \, n_d] \; : \; (i,j) = P_1^l \vee (i,j) = P_{r_l}^l \\ 0 \text{ otherwise} \end{cases} \qquad (22)$$

where as defined earlier, P_1^l and $P_{r_l}^l$ are the first and last points in the C^l edge line segment. It should be noted that the amount of data that has to be processed in F_d^p is considerably smaller than the amount of data that has to be processed in F_d^e, and so processing F_d^p is more efficient.

In the process of dashed lines reconstruction, a tube-directional dilation operator $D_{\theta, W, S}^L$ is applied on F_d^p, where the tube parameters are adapted in each point according to the component C^l to which the point belongs, and the strictness S is set to be minimal. Since edge line segments may be long and somewhat curved, for dashed lines reconstruction purposes, the characteristics Θ, and Δ that are used to set the parameters of the tube morphological operators, are determined according to an ending section of the edge line segment that is connected to the point in which the adaptation is required. Given an edge line segment C^l, its starting and ending sections with maximal length of k are defined by S_k^l and $\mathcal{E}_k^l$ respectively, where S_k^l is given by:

$$S_k^l \equiv \begin{cases} (P_k^l, \ldots, P_1^l) \text{ if } k < r_l \\ (P_{r_l}^l, \ldots, P_1^l) \text{ otherwise} \end{cases} \qquad (23)$$

and $\mathcal{E}_k^l$ is given by:

$$\mathcal{E}_k^l \equiv \begin{cases} (P_{r_l-k+1}^l, \ldots, P_{r_l}^l) \text{ if } k < r_l \\ (P_1^l, \ldots, P_{r_l}^l) \quad \text{otherwise} \end{cases} \qquad (24)$$

Therefore, given a point P in F_d^p that belongs to the component C^l, the tube parameters required for that point are based on $(\Lambda(C^l), \Theta(S_k^l), \Delta(S_k^l))$ if $P \in S_k^l$, or $(\Lambda(C^l), \Theta(\mathcal{E}_k^l), \Delta(\mathcal{E}_k^l))$ if $P \in \mathcal{E}_k^l$.

In the process of dashed lines reconstruction, all the points (i, j) for which $F_d^p(i, j) = 1$ are scanned. For each such point, a new image is created with the adaptive tube dilation of that point. This new image is then intersected with F_d^p. If the resulting image contains any points besides the point (i, j) itself, the distance from (i, j) to all the other points is evaluated, and the point which is the closest to (i, j) is selected as the point to which (i, j) should be connected to. The obtained connection is kept in a required connections list, and the two points are deleted from F_d^p. Since that process is repeated separately for each F_d^p plane, a preference is given to the connection of collinear dashes. A separate processing of each F_d^p plane, may lead to errors in which curved dashed lines are not connected in areas of high curvature, since the dashes in such areas reside in different edge planes. Therefore a concluding iteration is performed similarly on the union of all the remained unconnected points in $\bigcup_{d=0}^{3} F_d^p$.

Figure 5 presents an example for reconstruction of linear dashed lines, where Fig. 5-a presents the original dashed lines image, Fig. 5-b presents the union of four directional edge planes, Fig. 5-c presents results of dashed lines reconstruction obtained by the proposed approach, and Fig. 5-d presents results of dashed lines reconstruction without separation into directional edge planes. The gray rectangles in Fig. 5-b represent the adaptive tubes generated during the search process. As demonstrated, the results obtained without separation into directional edge planes are less natural than the results obtained with the separation into directional edge planes. Examining Fig. 5-b, it could be observed that the adaptive tubes that are used for the lines at angles that are multiplies of $45°$ are narrower than the tubes that are used for the other lines. The reason for more confidence in dashes which are sampled from discrete lines at angles that are multiplies of $45°$, is that such lines contain only one direction in their chain code [7], and so a single direction in the sampled dashes leads to a more accurate identification of their direction.

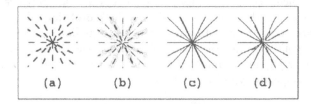

Fig. 5. Reconstruction of intersecting linear dashed lines

Figure 6 presents an example for reconstruction of curved dashed lines, where Fig. 6-a presents the original dashed lines image, Fig. 6-b presents the union of four directional edge planes and the tubes that are used during the reconstruction process, and Fig. 6-c presents the results of dashed lines reconstruction. As can be observed, the proposed approach manages to reconstruct intersecting curved dashed lines, even when they are close to other dashed lines. The small false connections in the restored image, are generated by the concluding iteration on the union of F_d^p. However, such false connections are limited by the width and length of the tube that is used, and thus can be removed by a post processing operation that gives preference to junctions of four lines over close junctions of three lines. Figure 7 presents an example for reconstruction of different dashed lines, where Fig. 7-a presents the original image, and Fig. 7-b presents the results of dashed lines reconstruction after the removal of character strings from the original image. As can be observed, even though the dashes in this image are unequally spaced, and have different lengths, the proposed algorithm still manages to reconstruct the dashed lines. It should be noted that in this example, the deletion of the character strings, cause one of the dashed lines to be broken by leaving a large gap in it. Such cases can be recovered by applying a post processing operation that connects restored dashed lines that cross a character string.

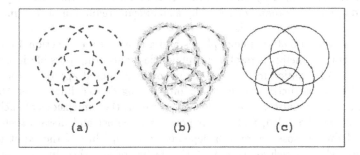

Fig. 6. Reconstruction of intersecting curved dashed lines

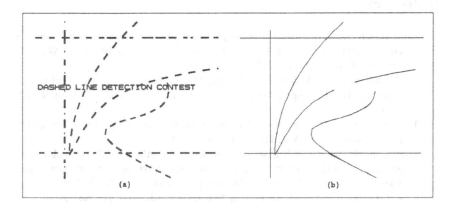

Fig. 7. Reconstruction of different dashed lines

6 Summary

An algorithm for detection and labeling of dashed lines in an image of dashes is presented. The algorithm uses adaptive directional morphological operators, where the adaptation is based on local characteristics of the data. The dashes image is first transformed into four directional edge planes, and dashes in each directional edge plane are associated with dashed lines by using the newly defined tube-directional morphological operators.

The usage of directional edge planes reduces the amount of data that has to be processed. In addition, it enables a more efficient implementation of algorithms, since the direction of the edge line segments in each directional edge plane is known in advance. In the process of reconstructing the dashed lines, the amount of data that has to be processed is further decreased, since only the edges of edge line segments are needed to be considered. The usage of adaptive tube-directional morphological operators limits the search space of possible dash connections to small zones determined by local characteristics. Therefore, the efficiency of the proposed approach is mainly derived by a significant reduction of the amount of data that has to be processed.

The usage of directional edge planes and adaptive tube-directional morphological operators, helps in resolving ambiguities concerning possible interpretations of the information in the image, and so the proposed approach manage to handle the reconstruction of close and intersecting dashed lines. Even though, in some cases, few false connections are generated, the dimensions of such false connections are bounded, and so it is possible to detect and discard them. The success of the proposed approach depends on the quality of the input image. The proposed approach assumes that the dashes of a dashed line are approximately directed in the direction of the dashed line. In cases where there are close dashed lines in which the dashes are twisted and do not follow the direction of the dashed lines, false connections will be generated.

References

1. G. Agam and I. Dinstein, "2-D Shape decomposition based on structures in a fuzzy relation matrix", in *Vision Geometry III*, R. A. Melter, A. Y. Wu eds., Proc. SPIE 2356, pp. 186-197, 1995.

2. D. Dori, Y. Liang, J. Dowell and I. Chai, "Sparse-pixel recognition of primitives in engineering drawings", *Machine Vision and Applications*, Vol. 6, pp. 69-82, 1993.

3. R. Kasturi, S. T. Bow, et al., "A system for interpretation of line drawings", *IEEE Trans. PAMI*, Vol. 12, No. 10, pp. 978-991, 1990.

4. H. Luo, G. Agam and I. Dinstein, "Directional mathematical morphology approach for line thinning and extraction of character strings from maps and line drawings", in *Proc. ICDAR'95*, Montreal, Canada, pp. 257-260, 1995.

5. V. Nagasamy and N. A. Langrana, "Engineering drawing processing and vectorization system", *Comput. Vision Graphics and Image Processing*, Vol. 49, pp. 379-397, 1990.

6. A. Pikaz and I. Dinstein, "Using simple decomposition for smoothing and feature detection of noisy digital curves", *IEEE Trans. PAMI*, Vol. 16, No. 8, pp. 808-813, 1994.

7. L. Wu, "On the chain code of a line", *IEEE Trans. PAMI*, Vol. 4, No. 3, pp. 347-353, 1982.

8. H. Yamada, K. Yamamoto and K. Hosokawa, "Directional mathematical morphology and reformalized hough transformation for the analysis of topographic maps", *IEEE Trans. PAMI*, Vol. 15, No. 4, pp. 380-387, 1993.

General Diagram-Recognition Methodologies

Dorothea Blostein

Computing and Information Science
Queen's University, Kingston Ontario, Canada, K7L 3N6
blostein@qucis.queensu.ca

The field of diagram recognition faces many challenges, including the great diversity in diagrammatic notations, and the presence of noise and ambiguity during the recognition process. To help address these problems, research is needed into methods for acquiring, representing, and exploiting notational conventions. We review several frameworks for diagram recognition: blackboard systems, schema-based systems, syntactic methods, and graph rewriting. Next we discuss the need for a computationally-relevant characterization of diagrammatic notations, the need to exploit soft constraints during diagram recognition, and the possibility that diagram generators may provide a useful source of information about notational conventions.

Keywords: diagram recognition, notational conventions, diagrammatic notations, contextual information, diagram classification, blackboard systems, schema-based systems, graph rewriting

1. Introduction

A variety of diagrammatic notations are widely used in society; examples include notations for music, math, architecture, and logic circuits. These diagrammatic languages use two-dimensional arrangements of symbols to transmit information. Understanding a diagram involves two activities: symbol recognition and symbol-arrangement analysis. Here we focus on the latter: analysis of the spatial arrangement of a group of symbols, relative to the given notational conventions of a 2D language, to recover the information content of a diagram. Important open questions in this area include:

- How should notational conventions be represented? We need systematic methods for representing conventions in a format suitable for computation. Current diagram recognition systems tend to use ad-hoc methods for representing notational conventions (e.g. [BlBa92] reviews music-notation understanding). Comprehensive expositions of notational conventions have been published (e.g. [Read79] for music notation); these present information in a style suited for human readers, but unsuited for computation. Where computationally-oriented descriptions of notations have been developed, most are oriented toward *generation* of the notation rather than *recognition* of the notation (e.g. [Rous88], [Knut79]).

- How can noise and uncertainty be handled? How should context be applied to reduce errors and uncertainty in segmentation and symbol-recognition?

- How can the great diversity of diagram types be handled? We need general, systematic methodologies for organizing recognition systems, representing notational conventions, and using contextual information to reduce errors. Ideally, such research would eventually result in a recognition system that is reconfigurable for various diagrammatic notations.

Much of the current research in diagram recognition is directed at recognizing some particular type of diagram. A comparison among recognition systems (Section 4) reveals common themes in the strategies used to design and implement recognition systems. We hope that over time this collective experience will give rise to a technology for general diagram recognition. An appealing analogy is provided by compiler technology, where general techniques for parsing and code generation greatly simplify the task of constructing compilers for new source and target languages. Diagram-recognition methods are difficult to generalize, due to the great diversity among diagrammatic notations, and due to the tremendous problems arising from noise and uncertainty in the input. Nevertheless, our research community as a whole would benefit from increased sharing of design and implementation strategies. Currently there are no criteria for choosing an appropriate recognition framework, or for determining how best to represent and apply notational conventions.

2. Six Diagram-Recognition Processes

Diagram recognition consists roughly of the following processes:

(1) early processing -- noise reduction, de-skewing, etc
(2) segmentation, to isolate symbols   } symbol recognition
(3) recognition of symbols

(4) identification of spatial relationships among symbols
(5) identification of logical relationships among symbols   } symbol-arrangement analysis
(6) construction of meaning

All of these processes require knowledge of notational conventions. For example, noise reduction methods (1) must preserve the presence of small or thin symbols, like decimal points in math and duration dots in music. Segmentation (2) needs information about how symbols overlap; an example is the separation of notes from staff-lines in music. Symbol recognition (3) needs information about symbol appearance; perhaps a font defining fixed symbols, and structural descriptions for parameterized symbols. The identification of spatial relationships (4) requires information about which spatial relationships are significant for encoding information. The identification of logical relationships among symbols (5), and construction of meaning (6), rely heavily on knowledge of notational conventions Further research is required to clarify how notational conventions should be represented and used.

World knowledge plays an important role in diagram recognition. For example, knowledge of disassembly and kinematics is used for the recognition of engineering drawings [VaTo94]. Similarly, human recognition of music notation uses the reader's knowledge of music theory and compositional styles. The acquisition, representation, and use of world knowledge is a broad and interesting topic, but one that lies beyond the scope of this paper.

While we find it useful to discuss diagram recognition in terms of the above six processes, the processes are not necessarily clearly delineated in an implementation, and they need not be performed in the indicated order. For example, partial identification of spatial and logical relationships can be performed prior to symbol recognition, as in math layout analysis [OkMi91]. After symbol identity is known, more detailed identification of symbol relationships is possible. (The identity of a symbol helps to determine which spatial and logical relationships are most meaningful. For example, in music notation the vertical location of a notehead or accidental conveys critical information, whereas the vertical location of a stem-end does not.)

Concurrency among the six recognition processes is possible, and can be useful in handling ambiguity, noise, and uncertainty. Two basic control strategies for handling uncertainty are:

- Sequential processing with lists of alternatives. All possible interpretations are carried along from one recognition stage to the next. For example, the symbol recognizer can produce a list of alternatives (including "noise") for each symbol it recognizes. During symbol-arrangement analysis, constraints are applied to eliminate alternative interpretations. Interleaving of constraint-application and information recovery may be required, since some constraints can only be applied once partial recognition results are obtained (e.g. [Fahm95]).

- Contextual feedback. Concurrent execution of recognition processes permits higher-level, contextual feedback to compensate for noisy input or to reject erroneous input. Blackboard systems, for example, provide such a control mechanism.

Sequential processing is used in [Balt94], with some discussion of limitations imposed by the need to carry forward all surviving alternatives. Sequential processing offers the advantage of isolating portions of the computation for intensive study. For example, Fahmy uses this approach to focus on symbol-arrangement analysis without the need to study symbol recognition [Fahm95]. A contextual feedback organization, on the other hand, offers increased flexibility in handling uncertainty, and may be necessary when the space of possible interpretations is extremely large. Further research is needed to develop general methods of providing contextual feedback.

3. Classes Of Diagram-Recognition Knowledge

The following classes of knowledge are present in a diagram-recognition system, either procedurally or declaratively. The implementation may mix these together, or may enforce a separation between them.

- Knowledge specific to the current diagram. A unified data structure can be used to represent the spectrum of interpretation stages, including the original image, partial interpretations, and the final interpretation. Practical advice about such a data structure is provided by [Balt94].

- Knowledge about the diagrammatic notation. The recognition system needs knowledge of notational conventions, but how best to represent notational conventions is very much an open question. Selected types of domain knowledge have well-studied data structures associated with them, such as dictionaries used in text recognition. The acquisition of domain knowledge is discussed in Section 6.

- Knowledge about diagram recognition. Generally-applicable operations include thresholding, line finding, and vectorization. Such general knowledge can form the kernel of a diagram recognition system, with the addition of domain-specific knowledge bases to configure the system for recognition of particular diagrammatic notations [Past94].

4. Frameworks For Diagram Recognition

Various system organizations and processing techniques have been proposed for general diagram recognition. In this section we survey several recognition frameworks, discussing how they have been applied to diagram recognition and document-image analysis. Frameworks that are discussed include blackboard systems, schema-based systems, syntactic methods, computational vision models, and graph rewriting. Sections 4.1 to 4.5 provide summaries of a large body of literature; readers not interested in details can choose to skip to the comparison of frameworks in Section 4.6.

4.1 Blackboard Systems

The blackboard architecture provides a general and flexible framework for combining diverse knowledge sources. (The name derives from an analogy to human experts communicating via a blackboard.) Here we discuss selected blackboard systems for document image analysis, including both text and diagram recognition. Existing blackboard-based recognition systems provide a wealth of experience in designing the detailed configuration of a blackboard system.

The three main parts of a blackboard system are a blackboard data structure (which contains recognition hypotheses), knowledge sources, and a control component [JaDB89]. Multiple, conflicting recognition hypotheses can be represented; knowledge sources commonly assign confidence values to the hypotheses. Authors mention various advantages of blackboard systems. These include the ability to integrate new knowledge sources into the system, the support for multiple hypotheses, the use of contextual feedback from high-level processing steps to low-level processing steps, and the ability to locally redo processing with new premises [Senn94]. Blackboard systems are able to handle both structure and randomness in the input, by acquiring evidence from diverse knowledge sources, without putting undo emphasis on any single source [WaSr89]. The control structure can invoke knowledge sources such that the effort expended in gathering evidence is in accordance with input complexity [WaSr89]. The blackboard architecture provides a flexible and adaptive order of processing; for example, clues can be opportunistically exploited to relate and interpret the text and diagram constituting a physics problem [NoBu93].

A blackboard is commonly divided into multiple levels of abstraction. Here we summarize the levels used in a sampling of systems. Kato and Inokuchi use five levels (image, primitive, symbol, meaning, goal) for music recognition [KaIn90a], and four levels (input diagram, symbol hypotheses, diagram hypotheses, recognition result) for circuit-diagram analysis [KaIn90b]. Vaxivière and Tombre use one level for each phase in engineering drawing analysis: lines and blocks, shafts, symmetric entities, functional setups [VaTo94]. Novak and Bulko use five levels (picture, text, picture-model, text-model, problem-model) to interpret the text and diagram defining a physics problem [NoBu93]. Sennhauser's blackboard for text recognition contains hypotheses related by a tree structure, with hypotheses at the page, block, chunk, and symbol level [Senn94].

Wang and Srihari use five levels (raw image, thresholded image, labeled image, text-line, and block levels) for address-block location in images of mail pieces; the main levels are divided into sublevels (seventeen in all) distinguishing e.g. machine-generated and hand-generated addresses [WaSr89]. The hierarchy of levels, models and algorithms used in text recognition are discussed in [Srih93]. Four main levels are used: the image level (divided into binary and grayscale levels), the character level, the linguistic level (divided into word, phrase, and sentence levels), and the textual structure level (divided into paragraph and document levels). Pasternak uses blackboard-driven building of aggregates in an adaptable drawing-interpretation kernel [Past94]; declarative geometrical constraints are applied to iteratively combine graphical objects into higher-level objects.

Knowledge sources can be organized in various ways. For example, Novak and Bulko suggest five types of knowledge-sources for symbol-arrangement analysis [NoBu93], which group related diagram elements, represent expectations about related objects, make inferences from common-sense rules and domain rules, relate diagram elements to *a priori* knowledge, and infer missing items.

Blackboard architectures provide a flexible control structure, which can be used to handle noise and uncertainty. In text recognition, character segmentation can be repeated if the classification results are unclear [Senn94]. A music recognition system can restore data to lower levels in the blackboard (for re-interpretation) if there is a contradiction at a higher level [KaIn90a]. This results in fast processing for clean input, with slower processing and more backtracking for noisy input. Similarly, in circuit-diagram recognition, the symbol hypotheses with the highest certainty factors are extracted first [KaIn90b]. Such processing allows cleaner parts of the input image to provide contextual information for interpreting noisier parts of the image.

Extensions to the basic blackboard system (knowledge sources, blackboard, and control component) are common. Here we summarize one such extension: Wang and Srihari's blackboard framework for object recognition in a highly variable set of images, including structured, partially structured, and unstructured images [WaSr89]. (The particular domain of interest is locating address blocks on mail pieces.) The six system components in this framework are a set of knowledge sources, the blackboard, a control mechanism (to select knowledge sources, combine evidence, update the focus-of-attention, and check termination), data for the control mechanism (a dependency graph that restricts the ordering of knowledge-source invocations, acceptance criteria for the object being recognized), a database of domain statistics (e.g. statistics about geometric features of logical blocks in a large sample of mail pieces), and a rule-based inference engine for interpreting the rules constituting a knowledge source. Each knowledge source contains rules for estimating the benefit and cost of using it, selecting parameters, and evaluating and interpreting results. Knowledge-source utility is estimated as a function of efficiency (fraction of knowledge-source-invocations that generate object candidates), effectiveness (fraction of generated candidates that turn out to be the desired object), average processing time, and the proportion of the input images to which the knowledge-source is designed to apply. The evidence generated by various knowledge-sources is combined using Dempster-Shafer's rules of combination, with the restriction that only singleton labels are assigned belief. Many aspects of this very interesting blackboard framework rely on the relatively simple structure of the object-recognition problem, but could perhaps be adapted to the diagram-recognition problem.

Blackboard architectures provide a means of coordinating fairly independent knowledge sources. This helps reduce the difficulties encountered in scaling up a recognition system. It is interesting to compare the scalability comments made by various authors. Sennhauser is quite optimistic about scaling up his blackboard system for text recognition [Senn94]. This system uses fairly regularly-structured, conceptually-simple knowledge sources, such as dictionaries and n-gram statistics; Sennhauser finds that the addition of these knowledge-source modules does not increase software complexity. Similarly, Wang and Srihari's blackboard framework isolates information about knowledge-source selection and utilization, so that changes to one knowledge-source do not cause side effects on other tools [WaSr89]. In contrast Vaxivière and Tombre are more cautionary in their discussion of scaling up a blackboard system for engineering drawing interpretation [VaTo94]. They use very high-level, irregularly-structured knowledge, including knowledge of disassembly and kinematics. Even their current, rather limited knowledge-base is somewhat difficult to manage, in maintaining consistency of rules, clarity, and an efficient search strategy. They conclude that their unstructured set of knowledge specialists is becoming unwieldy, and that the knowledge should follow a taxonomy, with hierarchical organization, perhaps as in a frame language.

4.2 Schema-based systems

Schema-based systems provide an alternative organization for diagram recognition. A schema class defines a prototypical drawing construct. During recognition a new schema instance is created to represent each drawing-construct that is found in the image. Two hierarchies are important in schema representations: the class hierarchy (to represent *specialization* relationships in knowledge organization) and the instance hierarchy (to represent *composition* of objects into more complex objects). Schema classes are organized into a tree-structured class hierarchy via the "subclass" relationship; this permits inheritance among schema-classes that describe related drawing constructs. Instantiated schemata are organized into an instance hierarchy via the "subpart" relationship; this provides the relationship between a schema and the smaller geometric entities that constitute it. Schemata can be used to represent both objects and relationships between objects, providing a uniformity of representation.

The Mapsee systems combine schema-based representations with constraint satisfaction, in order to interpret sketch maps [HaMa83] [MuMH88]. Schemata combine to form a *scene constraint graph*; constraints are propagated by a network consistency algorithm, to specialize the labels associated with schemata. Schema label sets have a specialization hierarchy, stating, for example, that both islands and mainland are landmasses, and both landmasses and waterbodies are geosystems. A constraint provides information such as "if a geosystem has a component river-system, road-system or mountain-range, then the geosystem-label is constrained to be a landmass". Mapsee-2 uses hypothesis trees to represent competing interpretations. Mapsee-3 instead uses labels (such as road, road/river, road/river/mountain) to represent competing interpretations; these labels are defined in a specialization hierarchy with multiple inheritance, so that specialization is not limited to a strict tree structure.

The Anon system [JoPr92] uses a three-layer structure (a strategy grammar, schemata, and an image analysis library) to recognize engineering drawings. At any given time, one schema is designated as the *controlling schema*. The controlling schema invokes the image analysis library, providing predictive information to guide the application of image analysis routines, and interpreting the results of the image analysis. The controlling schema encodes the image-analysis results as a token stream, which is passed up to the strategy grammar. The strategy grammar, an LR(1) grammar parsed by yacc, performs control actions (update the controlling schema, create a new schema, or designate a different schema to be the controlling one). Thus, spatial processing is directed by the controlling schema (where to look in the image, what to look for), whereas symbolic processing in managed by the strategy grammar (how to construct recognition hypotheses). The system is aimed at low and intermediate levels of diagram recognition: Anon analyzes an image to extract schemata representing drawing constructs. More global processes would be needed to integrate pieces of a drawing into a coherent whole. In contrast, the aforementioned Mapsee system does perform global processing (via constraint propagation), but does not perform segmentation and symbol recognition (dots and line-segments are provided as input).

4.3 Syntactic Methods

Syntactic methods can describe the structure of a diagrammatic notation. Domain knowledge is represented by a grammar, which consists of a start symbol and a set of rewrite rules. Depending on the formalism, a rewrite rule replaces one substring by another, one subtree by another, or one subgraph by another [Fu82]. This section considers the use of grammars for diagram recognition. Non-grammar-based use of rewrite rules is also possible (Section 4.5).

Anderson's grammar for the recognition of math notation provides an excellent case-study [Ande77]. The recursive nature of math notation is particularly well-suited for syntactic analysis. Anderson uses a set-based grammar (called a coordinate grammar). During top-down parsing, a production rule is given a set of symbols; the production rule applies spatial constraints to partition the symbol set into subsets, and assigns a syntactic goal to each subset. Anderson assumes the existence of a perfect symbol recognizer; recent work has eliminated this assumption [DiCM91]. Another syntactic approach to recognition of noisy math images uses stochastic grammars [Chou89]. For a more detailed review of mathematics-recognition, see [BlGr95].

Plex-grammars have been used to interpret dimensions in engineering drawings [CoTV93]. (A *plex* is an undirected graph; the nodes have named attachment points.) The parser mixes top-down and bottom-up processing. The processing of errors and uncertainty is delayed until after contextual information is obtained from higher-level analysis.

On a related note, a plex-grammar-based recognition system for three-dimensional objects is discussed in [LiFu89]. A 3D surface may be partially or totally occluded in the image plane. This makes it difficult to convert the input image into a three-dimensional plex to serve as input for the recognizer. A semantic-directed top-down backtrack recognizer is used, in which parsing directs the search for pattern primitives in the image.

4.4 Computational Vision Frameworks

Many of the problems we face in diagram recognition arise in general computational vision as well. Thus, frameworks formulated for computational vision are likely to contain ideas useful to us. Here we mention only one such framework. Joseph and Pridmore provide additional discussion of image understanding methods relevant to diagram recognition ([JoPr92], Section II).

Truvé describes a framework for high-level computational vision, based on multiple levels of interpretation, with feedback predictions [Truv90]. There are three phases to the process of going from one level of interpretation to the next: parsing, interpreting, and pruning (the *PIP paradigm*). The parsing phase uses local information to assign possible labels to image entities and groups of entities. The interpretation phase builds several interpretations, each assigning at most one label to a feature. The pruning phase uses global constraints to discard some interpretations. Parsing and pruning rules are expressed as multi-relational grammar rules (consisting of relations, constructors, and guards). The PIP paradigm applied to the sketch-map domain generates interpretations more efficiently than does Mapsee (Section 5.2). So far, Truvé has applied this framework to two relatively simple domains: sketch maps and the blocks world.

4.5 Graph Rewriting

In our experience, graph rewriting is a promising formalism for symbol-arrangement analysis. Graphs are a widely-used, flexible data structure, well-suited to the representation of document images. Graph-nodes represent entities in the physical image, or in the reconstructed logical representation. Graph-edges represent relationships among objects; these can range from physical relationships (e.g. adjacency or parallelism in the physical image), to logical relationships (e.g. a clef and a note in music notation, or matching brackets in math notation). Graph attributes, associated with nodes and edges, record non-structural information; initial attributes might record (x, y) image locations, whereas in the final graph, attributes record the meaning that has been extracted from the document image.

Graph rewriting can be used to transform an initial graph, derived from the original document image, into a final graph representing the logical structure and information content of the given document. Thus, the set of graph-rewriting rules encodes the *a priori* knowledge of the document type and its notational conventions.

Graph-rewriting rules can be organized in various ways: an unordered graph rewriting system, a graph grammar, an ordered graph-rewriting system (also called a programmed graph grammar), or an event-driven graph-rewriting system. These topics are discussed in the survey paper [BlFG95].

Ordered graph-rewriting has been used for various diagram-recognition applications, including circuit diagrams, music notation and math notation. In [Bunk82], ordered graph rewriting is proposed for the recognition of circuit diagrams. The high cost of parsing is avoided; instead ordered rule-application directly transforms an input graph (representing line-segment primitives in a circuit-diagram image) into an output graph (representing a circuit, with recognized components such as transistors and capacitors). Rule ordering is used, for example, to ensure that error-correcting rules are applied exhaustively (to close small gaps in lines and eliminate small overhanging lines) prior to the application of the component-recognition rules. This graph-rewriting approach is extended to music-notation recognition in [FaBl93]. The semantics of music notation depend both on nearby symbols (such as a sharp preceding a note) and on distant symbols (such as a distant key signature and a note). Thus, the necessary node-interactions cannot be predicted at the time of input-graph construction. Instead, a discrete input graph is used, and the necessary edges are built using Build, Weed, Incorporate phases of rewrite-rule application. (The Build phase creates edges between potentially-interacting nodes. The Weed phase deletes excess or conflicting associations. The Incorporate phase collapses the information from nodes and edges into the attributes of the remaining nodes.) This work is extended to address the uncertainty that may be associated with the identity of the notational primitives [Fahm95], and to perform recognition of mathematical notation [GrBl95]. We plan to investigate the use of graph rewriting in conjunction with a blackboard architecture. If data on the blackboard is stored as a graph, then selected knowledge sources could be coded as graph-rewrite rules.

4.6 Comparisons Among Recognition Frameworks

We have surveyed the use of several frameworks for diagram recognition: blackboard systems, schema-based systems, syntactic methods, computational vision models, and graph rewriting. These recognition frameworks make different provisions for handling various aspects of the recognition problem. Here we consider three aspects: representing levels of interpretation, processing noise and ambiguity, and representing knowledge about notational conventions.

Representing Levels of Interpretation

Levels of interpretation figure prominently in blackboard systems: recognition hypotheses on the blackboard are stored at different levels of interpretation. Schema-based systems use instance hierarchies to represent composition of objects into more complex objects. The class hierarchy represents specialization of knowledge about objects. Syntactic methods create levels of interpretation through the application of rewriting rules. Ordered graph rewriting similarly applies rewrite rules to create levels of interpretation; rule ordering offers explicit control over this process.

Processing Noise and Ambiguity

In a blackboard system, noise and ambiguity are processed by posting alternate recognition hypotheses on the blackboard, and eliminating inconsistent hypotheses as contextual information becomes available. This is a very general framework. The implementer is free to choose any search procedures and ambiguity-reduction algorithms; these are coded into the knowledge sources and the control component of the blackboard system. Schema-based systems add various mechanisms to handle noise and ambiguity. For example, the Mapsee system [MuMH88] uses schemata to construct a scene constraint graph; then constraints are propagated by a network consistency algorithm. The Anon system [JoPr92] uses a strategy grammar to direct schema-based construction of recognition hypotheses. Syntactic methods can handle noise through error-correcting parsing (typically a time-consuming operation for diagrams), or through stochastic grammars (as in [Chou89]). Ordered graph rewriting can handle noise in various ways, including explicit error-correcting production rules [Bunk82], or discrete relaxation [Fahm95].

Representing Knowledge about Notational Conventions

A central problem is representation of knowledge about notational conventions. In a blackboard system, this knowledge is stored in the knowledge sources; knowledge sources can contain an arbitrary mixture of declarative and procedural code. Schema-based systems use schema classes (with inheritance) to define notational conventions. Additional control mechanisms may be used as well. Syntactic methods store notational conventions in rewrite rules, which typically have associated attributes and attribute computations (e.g. [Ande77]). Ordered graph rewriting encodes notational conventions both in the graph rewrite rules, as well as in the ordering imposed on rule application.

Criteria for Choosing a Recognition Framework

We are not aware of criteria which can be used to choose the most appropriate recognition framework for a given application. Existing recognition systems illustrate that a general framework (e.g. blackboard system, schema-based system) can be adapted in many different ways. Unfortunately, it is difficult to determine the relative merits of the various adaptations. Also, the recognition frameworks are not mutually exclusive, but can be combined in various ways. For example, the general blackboard architecture could be combined with the use of schemata or graph rewriting. No matter what recognition framework is chosen, the creation of a diagram recognizer remains a difficult and time-consuming task.

This concludes our discussion of frameworks for diagram recognition. We now turn to two issues that are important to the advancement of diagram-recognition technology: the need for a computationally-relevant characterization of diagrams, and the need to exploit soft constraints during diagram recognition.

5. Computational Characterization Of Diagrams

It is difficult to formulate generalizations based on existing diagram-recognition research. One reason for this is the lack of a characterization or classification of diagram types. Without a characterization of diagrams, it is difficult to delineate the range of applicability of a particular diagram-recognition technique. Several different diagram characterizations could be computationally relevant:

- Classification of diagram types. Based on such a classification, a researcher could state that a recognition method applies to all diagrams of type "line drawing", "graph based", or whatever else. As an example, graph-like diagrams (including schematics, flowcharts, and piping diagrams) are given general treatment in the recognition system of [LSMS85].
- Classification of notational elements common to various diagram types. Researchers could then identify techniques to solve common subproblems. For example, many diagrams include lines (so "line detection" and "separation of lines and overlapping symbols" are general subproblems). Symmetry is important in some diagrams (the "symmetry detection" problem). In many notations, points or lines are labeled by placing alphanumeric strings nearby (the "label association" problem). Various notations use coordinate axes -- full coordinate systems for maps and y=f(x) plots, partial coordinate systems for pitch and duration in music -- making "coordinate system recognition and interpretation" a common problem. Various symbol-recognition problems can

be identified, some arising from notational primitives with fixed appearance (e.g. a note-head in music), others arising from parameterizable primitives (e.g. a beam or slur in music), and yet others arising from highly-complex, irregularly shaped primitives (e.g. lines in maps and y=f(x) plots).

Unfortunately, existing characterizations and classifications of diagrams are not computationally-oriented. For example, the recent article by Lohse et al. [LBWR94] provides many interesting examples of diagrams, and classifies these by asking human subjects to group together diagrams that seem similar. While the resulting classification is interesting, it is not directly useful for analyzing or generalizing diagram recognition methods. Another classification of diagrammatic notations is presented by Bertin [Bert83], but again not in a computationally-useful form.

Most diagrammatic notations are only semi-standardized, allowing many variations and drawing styles. For example, Tufte [Tuft83] describes numerous methods for displaying quantitative information, and encourages the designer of an illustration to be creative in the use of diagrammatic notations.

Computational characterizations of diagrammatic notations will emerge over time, as a result of ongoing work in the recognition of diverse diagrammatic notations. However, it is difficult to devise precise and impartial characterizations of notations based on experience with particular diagram recognizers: our impressions about diagram recognition often depend as much on the implementation of the given recognizers as on the inherent characteristics of the diagrammatic notation. Keeping these shortcomings in mind, we suggest a few dimensions along which notations could be characterized. We illustrate these ideas with comments based on our work in recognizing music and math notation [FaBl93] [Fahm95] [GrBl95].

- Subtlety of relative symbol placement. This characteristic of a notation affects how difficult it is to determine logical relationships among symbols from spatial relationships among symbols. In math notation, subtleties in the relative placement of symbols convey critical information. For example, minute changes in placement change our perception of whether two symbols are in a *subscript* or an *implied multiplication* relationship. In reading music notation, spatial relationships are less of a problem: staff lines provide a clear coordinate system for interpreting symbol placement. Thus, white-space plays an important role in the interpretation of math, more so than in music.

- Redundancy in the notation. Most diagrammatic notations have some redundancy in the representation of information. This provides a recognition system with constraints and cross-checks, particularly useful for the interpretation of noisy images. Although we cannot quantify this, it is our impression that music notation has more redundancy than does math notation. For example, image noise may make it difficult to determine whether a notehead is filled or unfilled; music notation offers redundant clues, such as "the number of beats in a measure matches the time signature", and "in multi-voice music, notes are placed so that beats line up across voices". Math notation offers less redundancy for resolving symbol-recognition ambiguities.

- Use of non-local symbol interactions. Both math and music notation rely on non-local symbol interactions, but in different ways. Math notation has a recursive structure, which results in relationships among symbol pairs -- such as

[and], or ∫ and dx -- that are separated by subexpressions. Due to this recursive structure, bottom-up processing of math notation can automatically construct many of the long-distance relationships. Math interpretation is complicated by the fact that symbols in many different positions can potentially affect the meaning of a given symbol; there is a dense web of interdependencies among symbols. Music notation, on the other hand, lacks a recursive structure, requiring more explicit processing of non-local symbol interactions (such as between a clef and a note, or between a sharp and all similarly-pitched notes following in the measure). The search for non-local symbol interactions is more constrained in music than in math; for example, only one direction must be searched to find the clef associated with a note.

The above comparisons between music and math notation are preliminary and imprecise. Interesting research issues are involved in improving these characterizations. Precise definitions are needed for terms such as *redundancy* or *non-local symbol interactions* in a notation. Means are needed to measure these quantities in a manner that is independent of a particular recognition-technique.

6. Acquiring And Exploiting Notational Conventions

Notational conventions supply us with the syntactic and semantic constraints needed to disambiguate and interpret a noisy document-image. Thus, the acquisition, representation and exploitation of notational conventions is central to diagram recognition.

6.1 What Can We Learn From Diagram *Generators*?

Computer processing of diagrammatic information includes recognition, editing, and generation of diagrams. Traditionally diagram recognition and generation systems have been written independently of one another. Current generation technology is more advanced than recognition technology -- many usable diagram generators are on the market (packaged with diagram editors), whereas diagram recognizers are in the research stage. Can diagram recognizers can be improved by exploiting the knowledge and experience embodied in diagram generators?

A recent communication-theoretic approach to document-image decoding uses the explicit modeling of document-image generation as a central part of a recognition system [KoCh94]. To control the search space, the image-generation models are limited to finite state machines. Although this results in simplistic image-generation models, impressive recognition results are obtained. Our current suggestion is to exploit far more sophisticated image-generation models, as are embodied by the notational conventions used in current commercial diagram-generators.

There is significant overlap in the notational conventions used for generation and recognition. However, complications arise due to the following differences in the two diagram processing tasks. First, diagram generation involves aesthetic decisions, knowledge of which are not crucial to recognition. Second, diagram recognition must deal with noise and uncertainty, problems which do not arise in generation. This is illustrated in Figure 1.

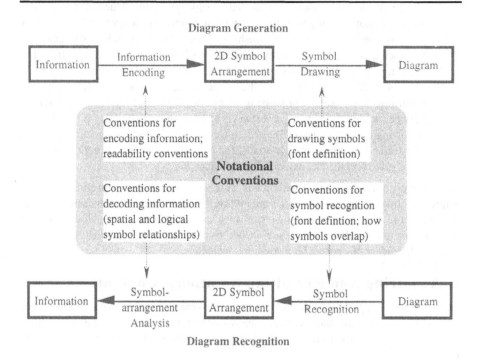

Figure 1 Diagram recognition and diagram generation systems both use notational conventions. For diagram generation, notational conventions dictate how to create an aesthetically pleasing diagram that encodes the given information. For diagram recognition, notational conventions dictate how to find logical relationships among symbols, and how to interpret these to recover the information content of a given diagram. (For simplicity, sequential processing is shown; in practice, diagram recognition can include feedback and concurrency between symbol recognition and symbol-arrangement analysis.)

Diagram-aesthetics are critical for human readability of a diagram, but current computer recognition systems do little to exploit these aesthetics. Ignorance of readability conventions simplifies the recognition task, but makes the recognizer less robust. In the following sections, we argue for the importance of exploiting readability conventions, and for the possibility that existing diagram generators are a valuable source of information about readability conventions.

6.2 Semantics and Readability: Hard and Soft Notational Conventions

Historically, diagrammatic languages arose as vehicles for communication, geared toward encoding information in a form that is easily readable by humans. For example, the lengthy evolution of music notation was shaped by the need to transmit pitch, duration, and phrasing information in a form that could be quickly read by performers [Slob81]. Thus notational conventions can be roughly characterized as being either *semantic conventions*, geared toward encoding information, or *readability conventions*, geared toward presenting the information in a form suited to human perception.

The distinction between semantic and readability conventions is fuzzy: any particular notational convention contributes both toward information-transmission and readability, but in varying proportions. Semantic conventions are small in number, and most of them can be described as local constraints on the diagrammatic notation. Readability conventions, which are sometimes referred to as aesthetic conventions, are large in number, difficult to state precisely, and complex to encode computationally. We illustrate this by considering notations for graphs and for music.

The diagrammatic notation for graphs has only few semantic conventions, such as: (1) graph structure is encoded by lines that connect nodes, and (2) node labels are written inside nodes. Readability conventions for graphs concern layout aspects such as limiting the number of edge crossings or edge bends, and spacing nodes appropriately.

The diagrammatic notation for music has a rich set of semantic and readability conventions [Read79]. Most semantic conventions are spatially localized, such as the use of noteheads, dots, flags, and beams to encode duration. An example of a non-local semantic convention is the propagation of accidentals from one note to a same-pitch note later in the measure. Numerous readability conventions govern the placement and spacing of music symbols. Of course, many notational conventions (such as "non-staff-line music symbols generally don't overlap") have important effects on both information-encoding and readability.

Two basic approaches are possible in using readability conventions for diagram recognition. The first is to simply ignore these conventions. This is the approach taken by most current diagram recognizers: they rely on hard semantic conventions and remain ignorant of readability conventions that are only "generally true". The second approach is to build knowledge of soft conventions into diagram recognizers. This provides the recognizer with much more contextual information for disambiguating noise and uncertainty in the input image. Since readability conventions strongly reinforce the information content of a diagram, a diagram-recognizer will be more robust if it is capable of interpreting the soft information provided by layout and spacing cues. These observations echo frequent comments in the diagram-recognition literature regarding the need to use context to resolve noise and ambiguity in the input image.

6.3 Sources of Information about Notational Conventions

Diagrammatic notations have complex notational conventions. Identifying and encoding these notational conventions is difficult and time-consuming. Currently, descriptions of notational conventions are available in the following forms.

- <u>Written descriptions oriented toward human use</u>: Comprehensive expositions of notational conventions have been published (e.g. [Read79] for music notation), but these present information in a style suited for human readers. (The typical reader of such a book is engaged in solving problems of typesetting the notation, not in solving problems of reading the notation.) Information is presented by example, with various notational conventions mixed together in each example. Unfortunately, no preconditions are stated for the rules and examples -- humans "use their judgment" in deciding when and how to apply the rules that are illustrated by example. This use of judgment is very difficult to mimic computationally.

- Written descriptions oriented toward computation: Where computationally-oriented descriptions of notations have been developed, most are oriented toward *generation* of the notation rather than *recognition* of the notation (e.g. [Rous88], [Knut79]).

- Algorithmic and declarative descriptions built into diagram-recognition systems: Current diagram recognition systems tend to use ad-hoc methods for representing notational conventions. The emphasis is on semantic constraints; few readability constraints are included.

- Algorithmic and declarative descriptions built into diagram-generation systems: Diagram generators contain comprehensive, though usually proprietary, descriptions of notational conventions. Much knowledge and experience is embodied in diagram-generation systems. For example, it is common for music-editing systems to be under development for a decade or more [Belk94]. Both semantic and readability constraints are included in diagram generators.

- Human experts: Interviews with a highly-experienced user of a diagrammatic notation can help in defining the notational conventions appropriate for a recognizer.

Existing descriptions of notational conventions provide valuable resources for the further development of diagram generators and recognizers. Given the large variety of diagrammatic notations [Bert83] [Tuft83] [LBWR94], it is desirable to develop general approaches to recognition and generation, applicable to whole classes of diagrams. Many interesting research problems are involved.

7. Conclusion

This paper has discussed the following research problems, aimed at furthering the emergence of a technology for general diagram recognition.

- Develop improved methods for acquiring, representing, and exploiting notational conventions. Notational conventions dictate how a given diagrammatic notation encodes information as a two-dimensional arrangement of symbols.
- Engage in comparative study of existing diagram-recognition frameworks. Here we briefly reviewed blackboard systems, schema-based systems, syntactic methods, and graph rewriting. Interesting commonalities and differences emerge when existing systems are studied.
- Develop a computationally-relevant characterization of diagrammatic notations. This will permit us to identify and exploit commonalities among diagrammatic notations.
- Develop methods to exploit soft constraints during diagram recognition. Many notational conventions concern diagram readability, and thus are observed "most of the time". These soft conventions provide important cues, but are unused by current diagram recognition systems.
- Consider diagram generators (built into diagram-editing systems) as a possible source of information about notational conventions, particularly about soft conventions.

Advances in the acquisition and exploitation of notational conventions can do much to further the state-of-the-art in diagram recognition.

References

[Ande77] R. Anderson, "Two Dimensional Mathematical Notation," in *Syntactic Pattern Recognition, Applications*, K. S. Fu editor, Springer 1977, pp. 147-177.

[BaIt94] H. Baird and D. Ittner, "Data Structures for Page Readers" *Proc. IAPR Workshop on Document Analysis Systems*, Kaiserslautern, Germany, Oct. 1994, pp. 323-334.

[Belk94] A. Belkin, "Macintosh Notation Software: Present and Future," *Computer Music Journal*, Vol. 18, No. 1, pp. 53-69, Spring 1994.

[Bert83] J. Bertin, *Semiology of Graphics: Diagrams, Networks, and Maps*, University of Wisconsin Press, 1983.

[BlBa92] D. Blostein and H. Baird, "A Critical Survey of Music Image Analysis," in *Structured Document Image Analysis*, Eds. H. Baird, H. Bunke, and K. Yamamoto, Springer Verlag, 1992, pp. 405-434.

[BlFG95] D. Blostein, H. Fahmy, and A. Grbavec, "Practical Use of Graph Rewriting," Technical Report No. 95-373, Computing and Information Science, Queen's University, January, 1995.

[BlGr95] D. Blostein, A. Grbavec, "Recognition of Mathematical Notation," in *Handbook of Character Recognition and Document Image Analysis*, Eds. H. Bunke and P. Wang, World Scientific, to appear.

[Bunk82] H. Bunke, "Attributed Programmed Graph Grammars and Their Application to Schematic Diagram Interpretation," *IEEE Trans. Pattern Analysis and Machine Intelligence* 4(6), pp. 574-582, Nov. 1982.

[Chou89] P. Chou, "Recognition of Equations Using a Two-Dimensional Stochastic Context-Free Grammar," *Proc. SPIE Visual Communications and Image Processing IV*, Philadelphia PA, pp. 852-863, Nov. 1989.

[CoTV93] S. Collin, K. Tombre, and P. Vaxiviere, "Don't Tell Mom I'm Doing Document Analysis; She Believes I'm in the Computer Vision Field," *Proc. Second Intl. Conf. on Document Analysis and Recognition*, Tsukuba, Japan, Oct. 1993, pp. 619-622.

[DiCM91] Y. Dimitriadis, J. Coronado, and C. de la Maza, "A New Interactive Mathematical Editor, Using On-line Handwritten Symbol Recognition, and Error Detection-Correction with an Attribute Grammar," in *Proc. First Intl. Conf. on Document Analysis and Recognition*, Saint Malo, France, September 1991, pp. 242-250.

[FaBl93] H. Fahmy and D. Blostein, "A Graph Grammar Programming Style for Recognition of Music Notation," *Machine Vision and Applications*, Vol. 6, No. 2, pp. 83-99, 1993.

[Fahm95] H. Fahmy, "Reasoning in the Presence of Uncertainty via Graph Rewriting," PhD Thesis, Computing and Information Science, Queen's University, March 1995. (TR 95-382)

[Fu82] K. S. Fu, *Syntactic Pattern Recognition and Applications*, Prentice Hall 1982.

[GrBl95] A. Grbavec and D. Blostein, "Mathematics Recognition Using Graph Rewriting," *Third International Conference on Document Analysis and Recognition*, Montreal, Canada, August 1995.

[HaMa83] W. Havens and A. Mackworth, "Representing Knowledge of the Visual World," *IEEE Computer*, October 1983, pp. 90-96.

[JaDB89] V. Jagannathan, R. Dodhiawala, L. Baum, Editors, *Blackboard Architectures and Applications*, Academic Press, 1989.

[JoPr92] S. Joseph and T. Pridmore, "Knowledge-Directed Interpretation of Mechanical Engineering Drawings," *IEEE PAMI*, Vol. 14, No. 9, Sept. 1992, pp. 928-940.

[KaIn90a] H. Kato and S. Inokuchi, "The Recognition System of Printed Piano Music using Musical Knowledge and Constraints," *Proc. IAPR Workshop on Syntactic and Structural Pattern Recognition.*, Murray Hill NJ, June 1990, pp. 231-248.

[KaIn90b] H. Kato and S. Inokuchi, "The Recognition Method for Roughly Hand-Drawn Logical Diagrams Based on Utilization of Multi-Layered Knowledge," *Proc. 10th Intl. Conf. on Pattern Recognition*, Atlantic City NJ, June 1990, pp. 443-473.

[Knut79] D. Knuth, "Mathematical Typography," *Bulletin of the American Mathematical Society*, Vol. 1, No. 2, March 1979.

[KoCh94] G. Kopec and P. Chou, "Document Image Decoding Using Markov Source Models," *IEEE Trans. Pattern Analysis and Machine Intelligence*, Vol. 16, No. 6, June 1994, pp. 602-617.

[LSMS85] X. Lin, S. Shimotsuji, M. Minoh, T. Saki, "Efficient Diagram Understanding with Chraracteristic Pattern Detection," *Computer Vision, Graphics, and Image Processing*, Vol. 30, 1985, pp. 84-106.

[LiFu89] W. Lin and K.S. Fu, "A Syntactic Approach to Three-Dimensional Object Recognition," *IEEE Trans. Systems Man and Cybernetics*, Vol. 16, No. 3, May 1986, pp. 405-422.

[LBWR94] G. Lohse, K. Biolsi, N. Walker, H. Ruetter, "A Classification of Visual Representations," *Communications of the ACM*, Vol. 37, No. 4, December 1994, pp. 36-49.

[MuMH88] I. Mulder, A. Mackworth, W. Havens, "Knowledge Structuring and Constraint Satisfaction: The Mapsee Approach," *IEEE Pattern Analysis and Machine Intelligence*, Vol. 10, No. 6, November 1988, pp. 866-879.

[NoBu93] G. Novak and W. Bulko, "Diagrams and Text as Computer Input," *J. Visual Languages and Computing*, Vol. 4, 1993, pp. 161-175.

[OkMi91] M. Okamoto and B. Miao, "Recognition of Mathematical Expressions by Using the Layout Structure of Symbols," in *Proc. First Intl. Conference on Document Analysis and Recognition*, Saint Malo, France, September 1991, pp. 242-250.

[Past94] B. Pasternak, "Processing Imprecise and Structural Distorted Line Drawings by and Adaptable Drawing Interpretation Kernel," *Proc. IAPR Workshop on Document Analysis Systems*, Kaiserslautern, Germany, Oct. 1994, pp. 349-363.

[Read79] G. Read, *Music Notation: A Manual of Modern Practice (Second Edition)*, Taplinger Publishing, New York, NY, 1979.

[Rous88] D. Roush, "Music Formatting Guidelines," Technical Report OSU-3/88-TR10, Department of Computer and Information Science, The Ohio State University, 1988.

[Senn94] R. Sennhauser, "Integration of Contextual Knowledge Sources Into a Blackboard-based Text Recognition System," *IAPR Workshop on Document Analysis Systems*, Kaiserslautern, Germany, Oct. 1994, pp. 211-228.

[Slob81] J. Sloboda, "The Uses of Space in Music Notation," *Visual Language*, Vol. XV, No. 1, pp. 86-112, 1981.

[Srih93] S. Srihari, "From Pixels to Paragraphs: the Use of Contextual Models in Text Recognition," *Proc. Second Intl. Conf. Document Analysis and Recognition*, Tsukuba, Japan, Oct. 1993, pp. 416-423.

[Truv90] S. Truvé, "Image Interpretation Using Multi-Relational Grammars," *Proc. Third International Conference on Computer Vision*, pp. 146-155, December 1990.

[Tuft83] E. Tufte, *The Visual Display of Quantitative Information*, Graphics Press, 1983.

[VaTo94] P. Vaxivière and K. Tombre, "Knowledge Organization and Interpretation Process in Engineering Drawing Interpretation," *Proc. IAPR Workshop on Document Analysis Systems*, Kaiserslautern, Germany, Oct. 1994, pp. 313-321.

[WaSr89] C. Wang and S. Srihari, "A Framework for Object Recognition in a Visually Complex Environment and its Application to Locating Address Blocks on Mail Pieces," *International Journal of Computer Vision*, Vol. 2, 1989, pp. 125-151.

Automatic Learning and Recognition of Graphical Symbols in Engineering Drawings

B. T. Messmer and H. Bunke

Institut für Informatik und angewandte Mathematik,
University of Bern, Neubrückstr. 10, CH-3012 Bern, Switzerland,
email: messmer@iam.unibe.ch, bunke@iam.unibe.ch

Abstract. In this paper, we propose a system for the recognition and the automatic learning of hand-drawn graphic symbols in engineering drawings. The graphic symbols and the drawings are represented by attributed relational graphs. The recognition process is formulated as a search process for error-tolerant subgraph isomorphisms from the symbol graphs to the drawing graph. In the beginning, there is a limited set of graphic symbols that are known to the system. The learning algorithm is able to identify new, i.e. unknown, symbols. From the set of all unknown symbols, representative candidates are selected and integrated into the database of known models. The system has been completely implemented and succesfully tested on a number of hand-drawn input pictures.

1 Introduction

In the past few years the automatic processing and analysis of graphical data has become the focus of intensive research. There are numerous applications in which the detection and recognition of graphic symbols is an important task [1, 2, 3, 4, 5]. However, often it is not sufficient to simply recognize symbols that are already known, but it is also desirable to automatically learn unknown symbols.

The discipline of machine learning is concerned with algorithms by which a system can automatically construct models, knowledge, or skills from the outside world. Generally, there are different categories of approaches to machine learning, like learning in neural networks, symbolic learning from examples, case-based learning, explanation-based learning, or reinforcement learning. For a general introduction to the area of machine learning and an overview, the reader is refered to [6, 7, 8].

While research on machine learning in the general artificial intelligence field has produced interesting results, machine learning in the domain of computer vision is still in its infancy. This situation is highly unsatisfactory because there are many problems in computer vision which will eventually be solved only if efficient learning schemes are available. For example, the manual construction of vision models is complicated, time consuming, and error prone. More seriously, perfect vision models can often not be built because the real world is unpredictable to some degree. Also, the type of noise or distortion that corrupt the pattern in reality is often only partly known. Therefore, it is impossible to design a computer vision or pattern recognition system that incorporates all models and knowledge that might be needed during runtime. Clearly, machine learning seems very promising to solve some of these problems. Overviews of the state of the art of machine learning in computer vision can be found in [9, 10, 11, 12].

In this paper, we propose a system, which combines pattern recognition techniques with machine learning concepts in order to solve the problem of recognizing and learning graphical symbols in engineering drawings. The drawings and the symbols they contain are represented by attributed relational graphs. This representation is translation, scale and rotation invariant. Symbols that are known to the system are stored in a database. Instances of known symbols are detected in the drawing by determining subgraph isomorphisms from the symbol graphs in the database to the graphs representing a complete input drawing. In order to detect distorted symbols as well, a set of error correcting edit operations has been defined. The graph matching component of our system is based on the network approach [13, 14, 15]. This approach allows a compact representation of the database of known symbols, and is very fast. The time complexity of the method is only sublinearly dependent on the size of the database. The learning component of the system is closely connected to the graph matching procedure. It makes use of the fact that the network approach supports incremental updating of the database of known symbols. I.e., new symbols can be added to the database during runtime without the need to recompile the database from scratch. Given an initial database of symbols, the system tries to recognize all the symbols in an input drawing. Detected symbols are removed from the drawing and the remaining line segments are grouped into possible new symbols according to heuristic rules. These rules encode the basic characteristics of the underlying class of symbols and are therefore strongly application dependent. Out of the set of possible new symbols, only some representatives, which are necessary in order to interpret the complete drawing, are learned and added to the database of symbols.

2 Symbolic Representation

In the system described in this paper, symbols and drawings are represented by attributed relational graphs. We allow that symbols appear at any position with varying size and orientation in a drawing. Therefore, the encoding of the symbols as attributed graphs must be translation, scale and rotation invariant. First, the drawings are given as scanned gray-level images. In a preprocessing phase, the images are binarized and thinned, and straight line segments are extracted. An example is shown in Figs. 1 and 2. For these steps, standard algorithms of image processing are applied. The current version of the system is restricted to drawings with straight line segments only, but it is straightforward to extend it so as to allow curved line segments, too.

Each line segment extracted during preprocessing is represented as a vertex of a graph. If two line segments l_1 and l_2 have an endpoint in common and

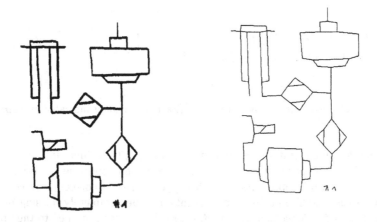

Fig. 1. Example of an input image after binarization.

Fig. 2. The image of Fig. 1 after thinning.

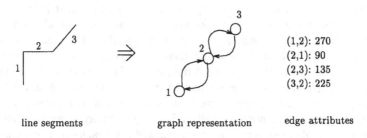

line segments   graph representation   edge attributes

(1,2): 270
(2,1): 90
(2,3): 135
(3,2): 225

Fig. 3. Graph representation of a simple line drawing.

v_1 and v_2 are their corresponding vertices, then the graph contains a directed edge (v_1, v_2) from v_1 to v_2 and a directed edge (v_2, v_1) from v_2 to v_1. There is an attribute associated with each of these edges. It denotes the angle that is enclosed between the two corresponding line segments. An example is shown in Fig. 3. Clearly, the proposed attributed graph representation is invariant with respect to translation, scaling and rotation.

Often, the drawings are corrupted by noise. Consequently, the graph representing a drawing will not always contain only exact instances of the symbol graphs. In order to correct local distortions in a drawing, we propose a number of graph edit operations. At run time, these edit operations will be applied to the graph representation of a drawing. In Fig. 4a a symbol is displayed; Figs. 4b-g show six distorted versions of it that illustrate the different types of errors that can be corrected by means of the edit operations. The first type of distortion affects the angles between two incident line segments (Fig. 4b). As a consequence, the attributes of the directed edges in the corresponding graph representation are not identical to the edge attributes in the original symbol. Therefore, these edge attributes must be subject to an *attribute substitution*. Secondly, line seg-

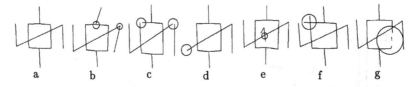

Fig. 4. Possible distortions in input drawings. (Each distortion is indicated by a circle.)

ments that are incident in the original symbol may not be incident in the drawing (Fig. 4c). For this reason, there will be no edges connecting the corresponding vertices in the graph representation. This type of error is corrected by *edge insertions* in the graph. Thirdly, line segments may be completely missing in the drawing (Fig. 4d). Consequently, a *vertex insertion* must be applied to the graph representation of the drawing. The edges connecting the new vertex to the rest of the graph need to be inserted, too. In the fourth case, a line segment may be erroneously intersected by some other line segment such that, instead of one vertex, there will be two vertices representing the line segment (Fig. 4e). In this case, a *vertex merge* must be performed to restore the original symbol. Fifth, additional line segments may appear at an intersection of original line segments (Fig. 4f). The graph of the drawing then contains additional vertices that are directly connected to the vertices of the symbol. For this type of distortion, no special edit operation is necessary because the matching component is based on subgraph isomorphism detection and consequently, additional vertices in the input graph are simply ignored. Finally, a line segment may be erroneously broken into several smaller segments that are separated from each other by empty space (Fig. 4g). In this case, the drawing graph will contain several additional vertices that are not connected to the rest of the graph via edges. To correct errors of this type, the graph matching algorithm performs a vertex insertion analogously to the operation performed for the distortion in Fig. 4d.

In order to model the fact that some of the distortions described above are more likely than others, each of the edit operations is associated with some cost. These costs are user-defined and must be empirically determined. The costs used in the application described in this paper are given in Section 5.

The process of recognizing a particular symbol in a drawing can be formulated as the search for an error-tolerant subgraph isomorphism between the symbol graphs and the drawing graph. The objective of the error-tolerant subgraph isomorphism search is to minimize the total costs of the edit operations that are necessary in order to find an exact subgraph isomorphism between the symbol graphs and the drawing graph. We call the minimal edit cost of an error-tolerant subgraph isomorphism from a graph g_1 to another graph g_2 the *graph distance* between g_1 and g_2 and denote it by $d(g_1, g_2)$. As the database of the known symbols usually contains more than one element, the recognition task for a given drawing consists of finding all symbol graphs for which the graph distance to the drawing graph is less than a certain threshold. For this purpose, we use an efficient graph matching algorithm that we describe in the next section. The

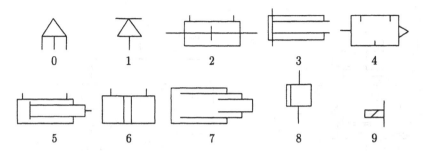

Fig. 5. Graphic symbols denoting machine parts. (The numbers are symbol identifiers.)

algorithm was developed by the authors independently of the present application [13, 14, 15].

3 Recognition Based on Subgraph Isomorphism

The algorithm for finding an error-tolerant subgraph isomorphism from a symbol to a drawing is based on a compact representation of the database of symbols. This representation is generated in an off-line compilation step. The basic idea of the compilation procedure is to find common substructures among the symbols, and to represent the symbols in terms of these substructures. For this purpose, the symbols are recursively split into smaller components. The smallest components taken into consideration are single line segments. For each of these components, a matching module is created, which performs the task of finding subgraph isomorphisms from the actual component to the drawing. If part of some symbol appears multiple times in the same or in different symbols, there will be only one matching module for this part. This results in a very compact representation of the set of symbols. Moreover, at runtime the search for subgraph isomorphisms for a part that occurs multiple times in one or in different symbols is done only once. Consequently, the subgraph isomorphism search is computationally very efficient.

The matching modules for the components of the symbols are organized in a network structure. At the top of this network is a matching module that represents the smallest possible part of any symbol, namely, a single straight line segment (of any orientation). Directly connected to the module on top of the network are matching modules which represent parts of symbols that consist of more than one straight line segment. These matching modules are in turn connected to modules which represent successively larger parts of symbols. Finally, at the bottom of the network there is a matching module for each of the symbols in the database. In Fig. 5 a database of symbols from mechanical engineering is shown [16]. Fig. 6 illustrates the network of matching modules for the symbols 0,1,2, and 3. (The full network representing all symbols $0, 1, \ldots, 9$ can't be shown because of limited space.) On top of the network there is a matching module coresponding to single straight line segments of any size and orientation.

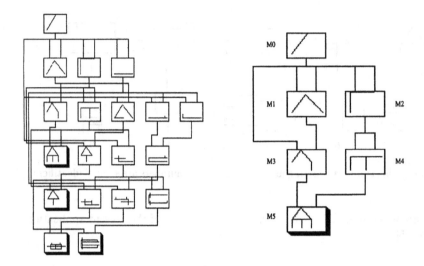

Fig. 6. The network representation for the symbols 0,1,2 and 3 of Fig. 5.

Fig. 7. The network for the symbol 0 of Fig. 5.

In the second level, there are three matching modules representing pairs of line segments that include an angle $\alpha_1 = 80°, \alpha_2 = 90°$ and $\alpha_3 = 180°$, respectively. The matching modules at the lower levels in the network represent parts of the symbols 0, 1, 2, and 3 that consist of three, four, ... line segments. Finally, the matching modules that are at the lowest level in the network – i.e. the terminal modules – correspond to the symbols 0,1,2, and 3. These modules are represented by shadowed rectangles in Fig. 6.

Note that a network like the one in Fig. 6 is automatically built from the graphs representing symbols like the ones shown in Fig. 5. A very important feature of our network compilation procedure is its incremental nature. Given a network N representing symbols $\{s_1, s_2, \ldots, s_n\}$, an extended network N' representing symbols $\{s_1, \ldots, s_n, s_{n+1}, \ldots, s_m\}$ can be built by just adding some more matching modules to N. The modules to be added represent those parts of $s_{n+1}, \ldots, s_m$ that are not yet represented by the network N. Particularly, there is no need to build N' from scratch, given N and $s_{n+1}, \ldots, s_m$. This feature of incremental extensibility is very important in the context of machine learning. If unknown symbols are discovered during the interpretation of a drawing, their symbolic representation can be easily and efficiently added to the existing network.

At runtime, all subgraph isomorphisms from a straight line segment to the input drawing are computed in the top module, and propagated to all direct successor modules. In each of these successor modules, the subgraph isomorphisms from the parent modules are combined to form subgraph isomorphisms for larger parts of the symbols. The propagation of the subgraph isomorphisms from one module to the next is controlled by the cost of the edit operations that are implied by the subgraph isomorphism. This process terminates when all subgraph isomorphisms with edit costs less than a certain threshold t_{recog} have been found. The matching modules for the symbols then contain all the subgraph isomorphisms with edit costs less than t_{recog}.

In order to illustrate the runtime behavior of the subgraph matching procedure, the part of the network of Fig. 6 that represents symbol 0 in Fig. 5 is shown in Fig. 7. In order to identify instances of symbol 0 in a drawing, all instances of straight line segments are recorded first by module M_0 on top of the network. Then two matching modules M_1 and M_2, are activated. M_1 extracts pairs of line segments enclosing an angle of $80°$, while M_2 checks for an angle of $90°$. The next module M_3, combines simple line segments from M_0 with instances recorded by M_1 to find a chain of three line segments, where the last two segments enclose an angle of $120°$. On the same level, the module M_4 inspects the result of M_2 and tries to combine any two instances found by M_2 into F-shaped configurations of line segments. Finally, module M_5 detects instances of symbol 0 by combining the results of M_3 and M_4.

Notice that the graph matching procedure described in this section is completely independent of the considered application. It is applicable to any problem domain where the objects under study can be represented by attributed graphs. The experiments reported in [13] demonstrated that the method is faster than other algorithms for error-tolerant subgraph isomorphism detection that have been described in the literature before. Particularly, the theoretical time complexity is only sublinear in the size of the database.

4 Learning New Symbols From a Drawing

The system starts with a number of a priori known symbols in the database. (This number may be zero.) Any input drawing is first interpreted with respect to this database of symbols. All line segments in the drawing that belong to recognized symbols are removed. The remaining line segments either constitute symbol interconnections or unknown symbols.

In order to identify unknown symbols, a few heuristic rules have to be defined. These rules correspond to a generic, meta-level definition of the symbols. In other words, they encode the basic characteristics of any graphical symbol in the application domain regardless of its particular shape. Of course, the characteristics of the symbols and the corresponding rules are application specific. For the present application, i.e. the set of symbols shown in Fig. 5, we defined three such rules: (1) The graph of any symbol is connected. (2) The graph of any symbol contains at least one, or possibly several adjacent closed loops of line segments. (3) Line segments that intersect a closed loop of a symbol belong to the same symbol. Note that all symbols in Fig. 5 comply with rules (1)-(3). Given a set of line segments that could not be associated to any known symbol, the identification of potentially new symbols is started by finding all closed loops of line segments. According to rule (2), a symbol must contain at least one closed loop. If two closed loops have one or more line segments in common, then it is assumed that they belong to the same candidate symbol. Finally, all line segments that intersect with a closed loop of a candidate symbol are assumed to be part of that symbol, according to rule (3). The identification of unknown symbols terminates when no more candidate symbols can be formed.

The procedure described in the last paragraph is able to extract unknown symbols from a drawing. Under the most primitive learning strategy, one could now integrate these unknown symbols into the network that represents the database of known symbols (see Fig. 6). However, a problem with this approach is that the network may grow unnecessarily large. An unnecessary growth of the database occurs if there are multiple instances of the same unknown symbol in a drawing, some of which have undergone some distortions and are represented, consequently, by different graphs. In order to overcome this problem, there is a second step in our learning procedure that aims at minimizing the number of symbols that must be stored in the database. This minimization is based on the detection of similarities between the unknown symbols. First, a subset of the set of unknown symbols is searched for such that for each unknown symbol there exists a symbol in the subset with a graph distance that is less than a certain threshold t_{learn}. For this purpose, all the unknown symbols are compared to each other by means of error-tolerant subgraph isomorphism. Let $C = \{c_1, c_2, \ldots, c_n\}$ denote the set of unknown symbols that have been extracted from the drawing. The graph distance from symbol c_i to c_j is given by $d(c_i, c_j)$. Note that normally $d(c_i, c_j) \neq d(c_j, c_i)$. We now apply a hierarchical clustering procedure to the set C of unknown symbols [17]. Initially, each candidate symbol c_i forms a simple cluster $U_i = \{c_i\}$. The distance of two clusters U_i, U_j is defined to be $D(U_i, U_j) = MAX\{MIN\{d(c, c'), d(c', c)\} | c \in U_i, c' \in U_j\}$. This means that we consider the minimal graph distances between any two symbols in the clusters U_i and U_j and choose the largest of these distances as the cluster distance. According to the distance measure D the two clusters with the least distance are now merged into a larger cluster. The distances between the new cluster and all the other clusters are computed and this process is continued until the distances of any two clusters exceed the threshold t_{learn}. At the end, each resulting cluster contains only candidate symbols with mutual distances that are less than the threshold t_{learn}. The size of the clusters is directly dependent on the value of t_{learn}. The larger t_{learn}, the more symbols will be merged into the same cluster and consequently less symbols will finally be learned. Out of each cluster U, a representative symbol is chosen such that the average graph distance from the representative symbol to all the other symbols in the cluster is minimized. That is, the symbol $c \in U$ for which the sum of the graph distances $\sum_{c' \in U, c' \neq c} d(c, c')$ is minimal is chosen as the representative symbol for the whole cluster U.

All the representative symbols found by the procedure described in the last paragraph are added to the database of known symbols and compiled into the network representation of the database. Note that the network grows less than linearly with each new symbol due to its ability of sharing common parts among different symbols. With the symbols that are automatically learned, it is possible to completely interpret any input drawing that contains graphical symbols that

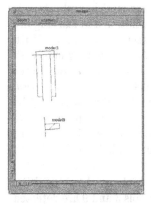 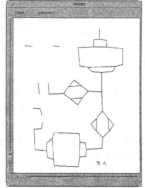 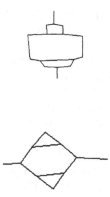

Fig. 8. Instances of symbols 3 and 7 detected in the first drawing.

Fig. 9. Fig. 2 after the symbols 3 and 7 have been removed.

Fig. 10. First and second symbol learned from the drawing in Fig. 3.

follow the rules (1)-(3). Furthermore, by learning the representative symbol from the "center" of each cluster of candidate symbols, an interpretation with minimal edit cost is guaranteed. In the following, we illustrate some of the details of the recognition and learning process by means of an example.

5 Experimental Results and Example

We illustrate the process of recognizing and learning symbols with an example dealing with machine layouts. Notice that our proposed procedure can be very easily adapted to other types of engineering drawings. Depending on the structure of the engineering drawings and the corresponding symbols it is necessary to empirically define the thresholds t_{recog} (Section 3) and t_{learn} (Section 4) and cost functions for the edit operations (Section 2). In our application, we defined the costs of the edit operations as follows (see also Fig. 4). We set the cost for substituting an angle α_1 with an angle α_2 to $(\alpha_1 - \alpha_2)^2$. (Note that for the cost calculation, the angles are given in radians instead of degree.) The cost for the insertion of an edge with an angle attribute α_1 between two non-intersecting line segments with distance d and enclosed angle α_2 was set to be $(\alpha_1 - \alpha_2)^2 * d$. The cost for inserting a vertex into the drawing graph was constantly set to 3. With this, we modeled the fact that substituting angles with a difference larger than $100°$ (1.74 radians) is highly unlikely and therefore a line segment should be inserted instead. Finally, the cost for merging two vertices that represent two collinear and touching line segments into a single vertex representing a straight line segment are constantly set to zero.

Based on these edit costs it is now possible to define the recognition threshold t_{recog} and the learning threshold t_{learn}. Because t_{recog} controls the degree of distortion that known symbols may have and t_{learn} defines the maximal distance among unknown symbols that are represented by the same model graph, there is a strong relationship between the two thresholds. If recognition and learning are to be performed on the same input drawing, it is best to set $t_{recog} = t_{learn}$. With this setting, and due to the fact that for all unknown symbols a new symbol with a graph distance of less than t_{learn} is learned, it is guarantueed that any input drawing can be completely interpreted. If some special task in the application requires that heavily distorted input drawings are interpreted, then t_{recog} may be increased empirically beyond the value of t_{learn}. In the present application, we set $t_{recog} = t_{learn} = 2$. With these thresholds and the costs of the edit operations mentioned above, it is possible, for example, to detect symbols for which two angle corrections of 60° (1 radian) are necessary.

Using the symbols shown in Fig. 5 as models and applying the recognition procedure to the drawing in Fig. 1, instances of the symbols 3 and 9 are detected and displayed at their correct locations; see Fig. 8. In order to be able to interpret the rest of the drawing, new symbols must be learned. For this purpose, the line segments that have been matched with the symbols 3 and 9 are removed from the drawing. The remaining line segments are given in Fig. 9. The learning scheme described in the previous section is then applied to these remaining line segments. In the first step of the learning procedure, four candidate symbols according to the rules stated in Section 4 are identified. Then these candidate symbols are clustered according to their mutual graph distances. As a result, two clusters each containing two candidate symbols are formed. In Fig. 10 the representative symbols that are chosen out of each cluster are displayed. These symbols are now compiled into the network representation of the other symbols in the database. With the new symbols added to the network it is possible to completely interpret the drawing in Fig. 1. Subsequently, the learned symbols can be detected in new input diagrams. In Fig. 11 (and Fig. 12) a drawing is given which contains distorted versions of (old) symbol 6 and one of the new symbols. Both symbols are succesfully detected by the system and located at the correct positions (see Fig. 13). The process of recognition and learning can be continued for any number of additional input drawings.

The system described in this paper was implemented in GNU C++ and runs on a SUN SPARC5 Workstation. It was tested on about 30 drawings with a complexity similar to those shown in Fig. 1 and 11, using the symbols shown in Fig. 5 and other types of symbols. With the thresholds properly set, recognition and learning worked flawless. The threshold adaption process is uncritical as there is a well-defined meaning of each threshold, and suitable threshold ranges can be determined by manual inspection. Also, the system performance is quite stable under threshold variations.

6 Conclusion

We have proposed a recognition and learning scheme for graphical symbols in engineering diagrams. It is based on a new and efficient error tolerant graph matching algorithm. This algorithm is used for the recognition of symbols, and

133

  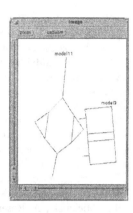

Fig. 11. Another drawing.

Fig. 12. Second drawing after thinning (containing distorted versions of old and new symbols).

Fig. 13. Detected and corrected symbols in second drawing.

in the learning procedure where possible symbols are clustered according to their mutual graph distance. Graph matching is based on a compact representation of the database, which can be incrementally updated and ensures a recognition time complexity sublinear in the size of the database. The learning and the recognition modules of the system work on general attributed graphs and can be easily adapted to other vision applications, particularly other types of engineering drawings.

Acknowledgement: This work was supported by the Swiss National Science Foundation under the Priority Program SPP IF. The authors want to thank Dr. D. Möri for making avaliable some low level image processing routines.

References

1. H. Bunke. Attributed programmed graph grammars and their application to schematic diagram interpretation. *IEEE Transactions on Pattern Analysis and Machine Intelligence PAMI*, 4(6):574–582, 1982.
2. B. Pasternak. Processing imprecise and structural distorted line drawings by an adaptable drawing interpretation kernel. In A. Dengel and L. Spitz, editors, *Proc. of IAPR Workshop on Document Analysis Systems*, pages 349–366, 1994.
3. A.H. Hamada. A new system for the analysis of schematic diagrams. *Proc. 2nd ICDAR,Tsukuba, Japan*, pages 369–372, 1993.
4. S.H. Kim and J.H. Kim. Automatic input of logic diagrams by recognizing loop-symbols and rectilinear connections. *Int. Journal of Pattern Recognition*, 8(5):1113–1129, 1994.
5. J.F Arias, C.P. Lai, S. Chandran, R. Kasturi, and A. Chhabra. Interpretation of telephone system manhole drawings. *Proc. 2nd ICDAR,Tsukuba, Japan*, pages 365–368, 1993.
6. A. Hutchinson. *Algorithmic Learning*. Oxford University Press, 1994.

7. J.W. Shavlick and T.G. Dietterich, editors. *Readings in Machine Learning*. Morgan Kaufman, San Mateo, 1990.

8. R. Michalski and G. Tecuci. *Machine Learning: A Multistrategy Approach*. Volume IV. Morgan Kaufmann Publishers, 1994.

9. Y. Kodratoff and S. Moscatelli. Machine learning for object recognition and scene analysis. *Int. Journal of Pattern Recognition and Artificial Intelligence*, 8(1):259–305, 1994.

10. B. Bhanu and T. Poggio (guest eds.). Special Section on Learning in Computer Vision. *IEEE Transactions on Pattern Analysis and Machine Intelligence PAMI*, 16(9):865–919, 1994.

11. S.R. Schwartz and B.W. Wah. Machine learning of computer vision algorithms. In T.Y. Young, editor, *Handbook of Pattern Recognition and Image Processing*, pages 319–359. Academic Press, 1994.

12. R.S. Michalski, A. Rosenfeld, and Y. Aloimonos. Machine Vision and Learning: Research Issues and Directions. *Proc. NSF/ARPA Workshop on Machine Learning and Vision, Harpers Ferry, West Virigina*, 1992.

13. B.T. Messmer and H. Bunke. Efficient error-tolerant subgraph isomorphism detection. In D. Dori and A. Bruckstein, editors, *Shape, Structure and Pattern Recognition*, pages 231–240. World Scientific Publ. Company, Singapore, 1995.

14. H. Bunke and B.T. Messmer. Similarity measures for structured representations. In M. M. Richter, S. Wess, K.-D. Althoff, and F. Maurer, editors, *Topics in Case-Based Reasoning, Lecture Notes on Artifical Intelligence*, pages 106–118. Springer Verlag, 1994.

15. H. Bunke and B.T. Messmer. Efficient attributed graph matching and its application to image analysis. In C. Braccini, L. DeFloriani, and G. Vernazza, editors, *Lecture Notes in Computer Science 974: Image Analysis and Processing*, pages 45–55. Springer Verlag, 1995.

16. *DIN Zeichnungswesen 1*, pages 269–286. Beuth, 1980.

17. B. S. Everitt. *Cluster analysis*. Edward Arnold, third edition, 1993.

A Hybrid System for Locating and Recognizing Low Level Graphic Items

F.Cesarini, M. Gori, S. Marinai, and G. Soda

Dipartimento di Sistemi e Informatica - Universtità di Firenze
Via S. Marta, 3 - 50139 Firenze - Italia
tel: + 39 55 4796361. fax: 39 55 4796363.
E-mail: {cesarini, marco, simone, giovanni }@mcculloch.ing.unifi.it

Abstract. This paper addresses the problem of locating and recognizing graphic items in document images. The proposed approach allows us to recognize such items also in the presence of high noise, scaling, and rotation. This is accomplished by a hybrid model which performs graphic item location by morphological operations and connected component analysis, and item recognition by a proper connectionist model. Some very promising experimental results are reported to support the proposed algorithms.

1 Introduction

A typical problem which arises in document analysis and recognition is that of locating and recognizing graphic items in a given area. When formulated in this way, the difficulty of the problem seems mainly related to the ambitious task of dealing with any graphic item, which imposes solutions that cannot rely on the item's structure. On the other hand, the graphic structure of today's documents is becoming increasingly sophisticated because of the presence of an impressive number of "unconventional" symbols. This makes the use of recognition algorithms tuned on specific graphic items very difficult. Obviously, when tuning a recognition algorithm on a specific graphic item, the recognition performance are likely to increase but, at the same time, developing such algorithms is a very time consuming process, mainly when dealing with many items. Moreover, there are cases in which the user wants the system to recognize new graphic items without asking for a program revision.

We suggest approaching these problems by using a hybrid system based on a connectionist model capable of recognizing graphic items on the basis of a learning from examples scheme and on a location algorithm which is based on morphological operations and connected component evaluation.

In particular, the item recognition is carried out by using an autoassociator-based architecture, where the input and the output layer have the same number of units. One hidden layer provides a compressed and nonlinear representation of the input information. The learning process consists of forcing the network to reproduce the input patterns to the outputs. For each item, a feedforward network acting as an autoassociator is trained. In order to simulate the behavior of the system in the real environment, the learning environment was composed of a collection of images obtained by corrupting the given item by artificial noise. The number of hidden units, as well as the other dimensions of the classifier, were established on the basis of a joint evaluation of convergence and generalization to new examples, simply by trial and error. A theoretical background on the use of autoassociators for detecting patterns can be found in [1, 2].

2 An Overview of the System

The system is based on a modular classifier for images of items of various kinds. The modularity is achieved by defining a neural classifier specialized in the recognition of each kind of items. New graphic items are recognized without changing previously defined classifiers. Each neural classifier is controlled by an algorithm that is charged of finding instances of the item and preprocessing the image according to the following scheme.

- *Item segmentation.*
  The aim of segmentation is to locate sets of pixels that are likely to represent a unique item [1].
- *Image preprocessing.*
  The portion of the incoming image, segmented at the previous step, is properly pre-processed, by obtaining a pattern that is essentially a compressed representation of the segmented item.
- *Recognition by autoassociator-based classifiers.*
  In the training phase, for each class the weights of the corresponding autoassociator are evaluated. The classification is made by determining the degree of matching with the pattern obtained at the previous step. The recognition is based on the Euclidean distance from input to output: the lower is the distance, the higher is the probability that the processed image contains the associated item.

Note that item segmentation and image preprocessing are performed only once for each item of the image. For each item, the autoassociator-based classification consists of using all the networks and assign to the unknown item the class of the auto-associator that generates lower error.

3 Item Segmentation

The aim of item segmentation is to identify sets of pixels that represent an item from a portion of the incoming image. General methods for item segmentation are difficult to develop, because there are differences in the visual appearance of general classes of items. The technique suggested in this paper works satisfactorily in different cases, provided that the distance between the connected

---

[1] For instance, when the item is a word, the segmentation is charged of separating all the words of the document.

components of the item is less than the distance of these components from parts of other items. Characters, words, and many types of logos reported in this papers are successfully located by using the proposed method.

Any word, for instance can be considered as a cluster of characters, such that the white gap between two neighbor characters is less than the gap between words, but the gap between consecutive rows of the text is smaller than the inter-character distance.

In order to merge the parts of an item, we combine morphological operations [3] (an optional dilation followed by an erosion) with connected component evaluation and dimensional check. The algorithm can be adapted quite easily to many situations, by changing the structuring element used by dilation. Fig. 1 shows examples of structuring elements used for locating words and components of a given electrical circuit. Fig. 3 displays the effect of the two operations when applied to a line of text, while Fig. 2 shows the steps of the location of components in an electric circuit. Basically, the item segmentation is carried out by performing an erosion or dilation of image, by detecting the connected component, and, finally, by a dimensional check.

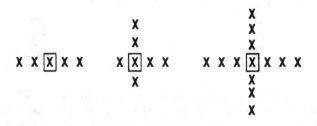

Fig. 1. Morphological processing. Examples of structuring elements: the first one is used for word detection (Fig. 3). The other structuring elements are used for erosion and dilation for the case of electrical circuit analysis (Fig. 2).

3.1 Morphological Operations and Connected Component Detection

The two basic morphological operations we use for item location are erosion and dilation. In erosion, the eroded image is a set containing all points in which the structuring element, "matches" the image points. In dilation, the dilated image is obtained by substituting the black pixels by the structuring element. The basic element of such operations is the structuring element (Fig. 1). The effect of erosion is to delete "small" objects in the image. The effect of dilation is to join together items composed of many parts (e.g. characters of a word).

A connected component is a set of black pixels such that each pair of pixels is connected by a path of pixels of the set. Classical algorithms [4] are used to detect these components into the image.

An example of such operation is given if Fig. 2. First, the smallest connected components corresponding to characters are removed from the original image. Second, the erosion operation gives an image where the wires are removed from the image. After that, an operation of dilation followed by connected component detection provides the image of Fig. 2.

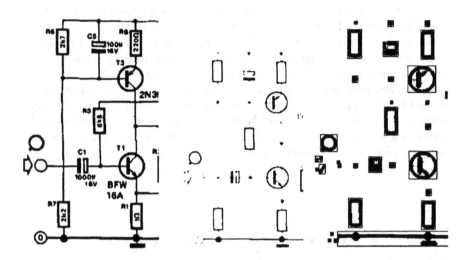

Fig. 2. An example of item location in an electrical circuit. The first is the original image. The second is the result after digit removal and morphological erosion. The last image is obtained by morphological dilation and connected components detection.

3.2 Dimensional Check

After the previous operations each item is expected to be contained into the upright rectangle enclosing a component. However, in general, there are some spurious components that have size or aspect ratio (ratio between horizontal and vertical item extensions) different from items that we have to recognize. A dimensional check is performed on the enclosing rectangles in order to eliminate these components. Depending on the graphic classification task, two checks are performed. If the system is designed to locate scaled items, then the aspect ratio is computed and is subsequently used to filter extracted items which do not match such ratio. If only fixed size items are considered, then we perform a dimensional check which filters those extracted items that are not size compatible.

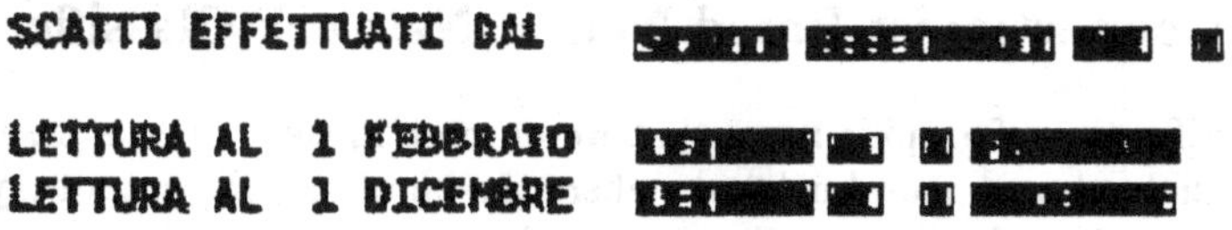

SCATTI EFFETTUATI DAL

LETTURA AL 1 FEBBRAIO

LETTURA AL 1 DICEMBRE

Fig. 3. An example of the processing carried out by dilation and connected components detection applied to word location.

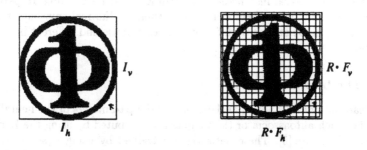

Fig. 4. A logo with the enclosing rectangle. On the right, the grid used to obtain the item representation.

4 Image Preprocessing

In the preprocessing step, the item is reduced to fit a fixed size frame. The frame size is chosen according to the architecture of the neural network, since the number of elements of the frame must be equal to the number of inputs of the network [2]. Given the frame, let F_h and F_v be the width an height respectively, and let I_h and I_v be the corresponding parameters for the item (Fig. 4). Reduction factors for horizontal and vertical directions are:

$$R_h = \lceil \frac{I_h}{F_h} \rceil, R_v = \lceil \frac{I_v}{F_v} \rceil \tag{1}$$

where $\lceil \cdot \rceil$ denotes the ceiling operation. In order to reduce distortions of the incoming item, the reduction rate R is chosen as the maximum between R_h and R_v. In order to compute the values that are introduced into the frame, an array of size $R \cdot F_h, R \cdot F_v$ is superimposed to the input item. The array is partitioned into disjoint square blocks of size $R \cdot R$, obtaining a grid superimposed to the item. For each of these blocks the number of black pixels is counted and divided by the total number of pixels into the block, R^2, in order to obtain the values (ranging from 0 to 1) that represent the item.

The size of these blocks, as well as the other dimensions of the classifier, is established on the basis of a joint evaluation of convergence and generalization to new examples, simply by trial and error.

---

[2] For logo recognition we used a frame size of 16 x 16.

5 A Connectionist-Based Neural Network Classifier

The classification of graphic items that we propose in this paper is based on an autoassociator-based neural network, where the input and the output layer have the same number of units (see Fig. 5). One hidden layer provides a compressed and nonlinear representation of the input information. The learning process consists of forcing the network to reproduce the input patterns to the outputs. Basically, for each kind of items, a feedforward network acting as an autoassociator is created, and several instances of the item are used for training. In particular, in order to simulate the behavior of the system in the real environment, these instances are associated with different degrees of noise and different instances of the item.

5.1 The Learning Phase

The learning phase is aimed at determining the parameters of the neural classifiers. For each autoassociator the weights are computed by using the Backpropagation algorithm [5]. These networks are trained by using positive examples only, that is examples of the class represented by each autoassociator. In some cases, such as for word recognition, some counterexamples were needed to improve the rejection capability of the autoassociators. In those case, the targets of the autoassociator for negative examples were simply forced to zero, with a significant improvement of performance. More sophisticated techniques for dealing with negative examples, however, are discussed in [2].

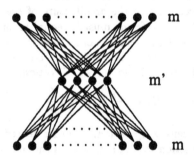

Fig. 5. A multilayer network acting as an autoassociator. The training forces the inputs, coming from positive examples, to reproduce the outputs.

5.2 The Recognition Phase

Let us denote by $X = \{x_j\}, j = 1, \ldots, m$ the vector that is obtained from the item location and preprocessing and that is subsequently used to feed the autoassociators. Let m and m' be the number of input and hidden units, respectively, being $m' < m$. Let $\{o_{ij}, i = 1, \ldots, N; \ j = 1, \ldots, m\}$ denotes the outputs of network A_i.

When feeding A_i with X, the correspondent error is $E_i(X) = \sqrt{\sum_{j=1}^{m} (x_j - o_{ij})^2}$. The classification of the unknown item X is carried out by using the following scheme:

1. *Auto-associators*
   Apply X to all N autoassociators, and compute error values: $\{\ E_i(X)\ \}$.
2. *Choose the class*
   For each pattern X, compute errors $\{E_i(X)\}$ of all N autoassociators. The class is chosen as the index for which the error is minimum.

While the training of the autoassociators is very expensive in terms of computational resources, their test is affordable even by ordinary workstations in real-time. Basically, the output of each networks takes $2 \cdot m \cdot m'$ multiplications, while the error takes m multiplications, that is we need $2 \cdot N \cdot m \cdot (m'+1)$ floating point operations.

Finally, Note that the proposed architecture is inherently modular, that is the introduction of a new item does not force us to re-train all the classifiers.

6 Experimental Results

In this section we give some experimental results for the application of the proposed method to word and logo recognition. These results have been obtained by introducing an artificial noise in original documents, which is described briefly in the following subsection.

6.1 Noise Generation

In order to simulate image defects of actual documents and to produce significant item data bases for training the classifiers, several models of noise have recently been suggested in the literature. Global and local document image defect models have been proposed. Global models [6] describe perspective dirstorsions occurring while scanning thick documents. Local models [6], [7] describe degradations due to perturbation in the printing, scanning and digitization process. In this paper we consider local models only. We developed a program, which is related to Baird's model [7], for generating noisy images to be used for both the training and recognition process.

Given a grey level image and a set of model parameters, the noise image generator computes a degraded bilevel image, operating with a sequence of operations.

- The rotation of the image is applied on the grey level image. This step simulates the "noise" that is due to paper positioning variations of the original document in the scanner bed (see e.g. *skew* in [7]).

Fig. 6. Examples of logos. In the used database, they are referred to as logo 4, logo 21, logo 20 and logo 24, respectively.

- The noise due to translation is simulated according to [8].
- The document is scaled in order to reach the desired final resolution.
- The optical distortions added by the scanner are modeled by a blurring of the image. According to [7], the point-spread function of the combined printing and imaging process is modeled as a circularly symmetric Gaussian filter with a standard error of *blur* in units of output pixel size.
- Finally, the binarization of the image is performed by means of a proper threshold (*thrs*). At this stage a sensitivity variation of pixel sensors is simulated.

Examples of the effects of this noise on logos are shown in Fig. 7, where the following parameters are used: *skew* = [-5, 5] (uniformly distributed in the range -5 degrees, +5 degrees), *blur* = $N(1, 0.3)$ (where $N(x,y)$ denotes a random valued normally distributed with mean x, and variance 0.3), *thrs* = [96,160] (in the image 0 is white and 255 black).

6.2 Logo Recognition

Some experiments were carried out concerning the problem of logo recognition, that is becoming increasingly relevant in document reading systems.

An interesting approach to the solution of such problem has been proposed in [11]. It is based on global invariants in order to prune the logo database, and local affine invariants to obtain a more refined match.

In this paper we show that auto-associators based classifiers perform very successfully in the problem of recognizing noisy and rotated images of logos. The database used in these experiments is available from the *Document Processing Group, Center for Automation Research, University of Maryland* and is the same used in the experiments presented in [12]. In our experiments we used a subset of this database, which is composed of the first 28 logos.

Rotated Logos. For each logo an autoassociator-based classifier was constructed. The number of neurons in the hidden layer was chosen by simple trial and error. The final architecture contained 256 inputs and outputs and 25 hidden units. The network was trained by 36 patterns, each of which represented a rotated instance of a logo. In particular, we presented to the network images that were rotated with steps of $\pi/18$.

The test set was created by considering the same 28 logos and generating, for each of them, 360 rotated instances, with steps of $\pi/180$. For each rotated logo the pattern was presented to all the 28 networks and the input-output errors were evaluated. We get a glimpse of the experimental results that we found by looking at Table 1, where the errors are reported for two logos, namely logos no. 3 and no. 24. These errors refer to the network associated with a given logo, when feeding it by another one which belongs to the test set. Since the same logo is fed by different rotations, the maximum and minimum error are reported in order to appreciate the networks behavior corresponding to logo rotations.

Logo 3 is a typical example, while logo 24 represents the worse result that we found on all the 28 logos.

Noisy Logos. In order to simulate actual noise, we used the program described in section 6.1. The parameters that we employed were: *skew* = [-5, 5], *blur* = $N(1, 0.3)$, *thrs* = [96,160].

Fig. 7. Different levels of noise added to logo 4 of Fig. 6.

Like for the previous case, for each logo an autoassociator-based classifier was constructed. This network had 256 inputs, 256 outputs, and 25 hidden units, and was trained by examples with an increasing degree of noise. Some noisy logos are shown in Fig. 7. Each training set was composed of 100 images.

The test set was created by considering the same 28 logos and generating, for each of them, 100 noisy logos with same kind of noise of the learning set.

The preprocessed pattern associated with each noisy logo was presented to all the 28 autoassociators and the input-output errors were consequently evaluated. In Table 1 the errors are reported for two logos, namely logos no. 15 and no. 21.

Like in the previous case, logo 15 is a typical example, while logo 21 represents the worse result that we found on all the 28 logos.

As we can see, in all but one case the maximum error for the correct class is less than the minimum error for any other class.

	Logo3		Logo24			Logo15		Logo21	
	min	max	min	max		min	max	min	max
Logo2	7.3	7.8	5.8	4.8	Logo2	5.4	6.6	6.1	6.7
Logo3	2.1	2.7	7.0	6.6	Logo3	5.9	6.9	6.1	7.0
Logo4	6.0	6.8	4.4	7.1	Logo4	6.2	8.4	5.6	7.5
Logo5	5.4	5.9	3.4	4.0	Logo5	7.0	8.9	6.1	8.1
Logo6	6.3	6.7	4.2	5.7	Logo6	4.3	5.4	5.2	6.7
Logo7	5.6	7.0	7.0	9.9	Logo7	5.7	6.6	5.7	6.5
Logo8	6.7	7.1	4.0	5.2	Logo8	6.8	8.3	6.5	7.3
Logo9	6.6	7.1	3.8	4.6	Logo9	8.0	9.5	7.1	8.4
Logo10	5.8	6.9	3.5	3.9	Logo10	6.3	8.1	6.9	7.9
Logo11	5.5	5.7	4.0	4.6	Logo11	2.9	3.2	3.9	4.6
Logo12	6.8	7.4	4.9	6.3	Logo12	5.4	6.9	5.5	6.8
Logo13	6.3	7.9	5.0	9.0	Logo13	6.8	8.5	6.4	7.1
Logo14	6.4	7.4	4.5	9.1	Logo14	7.3	8.7	6.0	7.2
Logo15	6.8	7.1	4.0	5.0	Logo15	0.5	2.9	7.2	5.0
Logo16	6.0	7.0	2.6	4.1	Logo16	4.1	5.6	5.8	6.8
Logo17	5.1	5.6	5.0	5.5	Logo17	3.9	6.0	3.9	5.7
Logo18	5.9	7.0	4.9	6.6	Logo18	4.8	6.1	6.0	6.6
Logo19	4.9	5.4	3.4	3.8	Logo19	5.9	7.5	5.7	6.9
Logo20	6.6	7.1	1.9	3.1	Logo20	8.4	10.0	7.2	8.5
Logo21	6.7	7.3	5.0	6.8	Logo21	6.4	7.8	0.4	1.0
Logo22	6.1	6.6	6.0	6.6	Logo22	5.2	6.6	6.4	7.3
Logo23	5.1	6.1	3.5	4.7	Logo23	5.6	6.9	6.0	6.8
Logo24	6.3	7.0	1.0	2.0	Logo24	8.7	10.7	7.6	9.7
Logo25	5.6	6.0	4.9	5.4	Logo25	5.9	3.5	4.9	6.3
Logo26	4.5	5.9	3.3	4.7	Logo26	4.3	5.4	5.3	6.2
Logo27	6.0	6.5	6.2	7.0	Logo27	5.3	6.5	6.2	7.0
Logo28	4.3	4.6	5.0	5.4	Logo28	5.2	7.4	5.3	7.2
Logo29	4.7	5.7	3.5	4.6	Logo29	5.7	6.4	6.4	7.3

Table 1. Experimental results for logos: the rows are associated with the trained networks, while the columns represent the logos to be recognized. The first table describes the results for rotated logos, while the second one describes the results for noisy logos.

6.3 Word recognition

We give some experimental results on the use of the system for locating one word from special forms used for the accounting of electrical services provided by Italian Company ENEL [3]. The recognition of these words is performed without needing the segmentation into characters. Such an approach to word detection turns out to be interesting when dealing with highly noisy forms, as in that case there is no significant difference between recognizing words and other graphic symbols. The motivation for approaching this task can be found in [9], where the general architecture of a form reading system is also described.

The word recognition has been extensively discussed in the literature by different approaches (see e.g.:[10]). Basically, different classes of algorithms arise depending on whether or not the word is segmented into single characters. The segmentation in characters turns out to be the only significant way of recognizing words of a very large dictionary, whereas very effective algorithms can be used for the recognition of words of small dictionaries when using the whole word as input to the classifier.

The experiments for word recognition were performed by simulating the actual noise by *DISTORT*, a program developed at MITEK Systems Inc. and available via Internet, that generates blurring noise on the basis of the selection of two parameters: the number of passes, and the probability of blurring.

The learning set was composed of 215 examples; 200 of them were used as positive examples and were obtained by noisy corruption of the word to be found. The noise was one pass of blurring with probability 0.4. The other 15 images were used as negative examples in order to improve the rejection capability of the autoassociators. In that case, the targets of the autoassociators were forced to zero.

Type of noise	number of tests	number of FP	number of FN
Blurring	1200	2	0
Speckle	400	0	0
Blurring + Speckle	1200	5	2
Total	2800	7	2

Table 2. Experimental results for noisy words location. Rows indicate different conditions of noise. FP and FN indicate, false positive and false negative, respectively.

Table 2 reports the experimental results we obtained. Four different forms were used and the system was asked to detect the requested word in different

---

[3] Ente Nazionale per l'Energia Elettrica.

noisy conditions as indicated in the table. Blurring, eventually added to speckle noise was used as artificial noise for corrupting the forms. The parameters of the blurring where from one to three passes with probability 0.2. The speckle noise consists of adding small black "spots" to the images with probability 0.01. The percentage of spots of size 1, 2 and 3 was 70 %, 20 %, and 10 %, respectively.

7 Conclusions

In this paper we have proposed a hybrid approach for the location and recognition of low level graphic items. While the location of these items is based on morphological operations and detection of connected components, the classification relies on a society of special neural networks, called auto-associators, where each network represents the model of a given graphic item.

Our preliminary experiments rely on simple trial and error for adjusting the optimal architecture, but the results are very promising in both the case of noisy and rotated graphic logo, and significant results have also been obtained for word recognition. The system is inherently modular and, moreover, the architecture based on autoassociators allows us to give reliable confidence of the recognized patterns [1, 2].

A limitation of the proposed approach is that it can hardly deal with cases in which the graphic items are very similar and, consequently the scaling up is critical for very large logo data base. In those cases a possible solution could be that of using proper multilayer perceptrons specifically trained to discriminate "confused" items, acting on the best logo candidates found by the autoassociators.

Our future research is essentially in two directions. First, we would like to explore algorithms for adapting the architecture also in terms of the number of connections (e.g. pruning algorithms), in order to improve the generalization to new examples. Second, the property that autoassociators have to provide reliable recognition scores, suggests us to include a sort of hypothesize and verify paradigm into the system, which are related to the location and recognition of graphic items, respectively.

References

1. M. Bianchini, P. Frasconi, and M. Gori. Learning in multilayered networks used as autoassociators. *IEEE Transactions on Neural Networks*, 6(2):pages 512 – 515, 1995.
2. M. Gori, L. Lastrucci, and G. Soda. Autoassociator-based models for speaker verification. *Pattern Recognition Letters*, To Appear.
3. J. Serra. *Image Analysis and Mathematical Morphology*. Academic Press, London, U.K., 1982.
4. D. H. Ballard and C. M. Brown. *Computer Vision*. Prentice Hall, Englewood Cliffs, N.J., 1982.
5. D. E. Rumelhart, G. E. Hinton, and R.J. Williams. Learning representation by error backpropagation. In *Parallel Distributed Processing*. MIT Press, 1990.

6. T. Kanungo, R.M. Haralick, and I. Phillips. Global and local document degradation models. In *Proceedings of the International Conference on Document Analysis and Recognition*, pages 730 – 734. IEEE Computer Society Press, 1993.

7. H.S. Baird. Document image defect models. In *Structured Document Image Analysis*, pages 547–555. Springer-Verlag, 1992.

8. H.S. Baird. Calibration of document image defect models. In *Symposium on Document Analysis and Information Retrieval*, pages 1 – 16, 1993.

9. F. Cesarini, M. Gori, S. Marinai, and G. Soda. A system for data extraction from forms of known class. In *Proceedings of the International Conference on Document Analysis and Recognition*, pages 1136–1140, 1995.

10. T.K. Ho, J.J. Hull, and S.N. Srihari. A computational model for recognition of multifont word images. *Machine Vision and Applications*, 6(6):157–168, 1993.

11. D.S. Doermann and A. Rosenfeld. The processing of form documents. In *Proceedings of the International Conference on Document Analysis and Recognition*, pages 497–501, 1993.

12. D. S. Doerman, E. Rivlin, and I. Weiss. Logo recognition. In *Center for Automation Research, Technical Report 3145*, 1993.

Automatic Interpretation of
Chemical Structure Diagrams[†]

Joe R. McDaniel* and Jason R. Balmuth
PSI INTERNATIONAL, Inc.
810 Gleneagles Court, Suite 300
Baltimore, Maryland 21286
joe@psiint.com

Abstract. Chemical structure diagrams, just as in engineering drawings, maps, and other technical diagrams, consist of solid and dashed lines (bonds), characters (atom symbols), and other symbols such as brackets, parentheses, wedges (stereo-up bonds) or dashed wedges (stereo-down bonds). In addition to recognizing these low-level elements of such drawings, other artifacts may be present — bonds intersections may be crossings or atom nodes, character strings may represent underlying chemical structure, and circles are sometimes used to represent ring-alternating bonding — requiring a considerable knowledge base of chemistry to be able to interpret correctly. This paper discusses the general processes used in the program Kekulé[1] that embodies this interpretation ability with more detailed explanations of how some problems relating to polygon approximation, dashed line and dashed wedge finding, and optical character recognition were solved.

1. Introduction

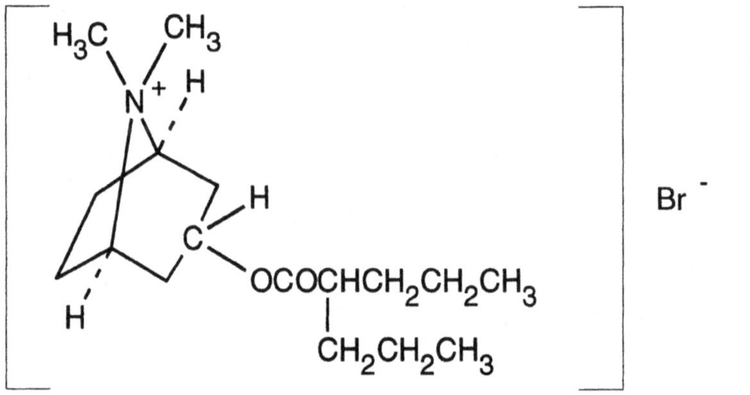

Fig. 1: **Typical chemical structure diagram.**

Chemists, like engineers and mapmakers, communicate information about the structure of chemical compounds via structural diagrams. While such diagrams work very well for communicating between people, computers cannot, in general,

[†] Work on this project was supported in part by the National Cancer Institute under SBIR Grant 5 R44 CA47241.
[1] Named for Friedrich August Kekulé von Stradonitz (1829-1896), who invented chemical structure notation as well as elucidated the structure of benzene.

read these diagrams with similar ease. What is needed is an equivalent to optical character recognition–optical chemical structure recognition that can automatically turn a structural diagram into a graph node and bond table or connection table that is understandable by computers.

Fig. 1 illustrates many of the notational conventions in chemistry that must be interpreted. The solid lines represent single bonds. The dashed lines are stereo-down bonds (wedges, not shown, are used for stereo-up bonds). The nitrogen and bromine (N^+, BR^-) have charges associated with each atom. The character strings (group formulas) H_3C, CH_3, $OCOCHCH_2CH_2CH_3$, and $CH_2CH_2CH_3$ all represent struture data in a shorthand notation with the additional need to recognize the bonding between atoms within the string to outside nodes or even atoms inside other group formulas. The bond from the nitrogen to the lower-left intersection (carbon is assumed if not explicitly shown) crosses another bond. The brackets are used for grouping (in this case to show a distributed charge). All of this notation (and most other common notation) is automatically recognized by Kekulé.

Interpreting such diagrams from scanned images of machine printed structures typically takes Kekulé from about 7 to 30 seconds depending on drawing complexity and the number of characters present. The alternative is to redraw structures manually, but this requires about 6 to 10 minutes per structure. In addition to its unique interpretation feature, Kekulé provides a full editor for manual drawing and editing of interpreted results. Even redrawing relatively simple structures is time-consuming and error-prone compared to processing scanned images automatically. Kekulé is the first successful attempt[2] [1, 2] to integrate all of the required elements of image processing, optical character recognition, structure editing, communication, and publication-quality output.

A scanner is used to capture a printed structural diagram. Kekulé then processes this scanned image to extract information on characters (atom symbols, charges, masses, repeat counts, etc.), lines (single, double, and triple bonds; brackets, etc.), and other objects (e.g., stereo bonds). This extracted information is then assembled into a connection table format. If errors are detected due to chemically unrecognizable atomic symbols or group formulas, the user is prompted for verification or correction. When interpretation completes, the user may use the editing features of Kekulé for tasks as simple as repositioning nodes or as complex as creating complete new structures. The resulting interpreted diagram can be conveyed in popular connection table formats — ISIS, MOLfile, ChemDraw, ChemIntosh/ChemWindows, SMILES, or ROSDAL format to chemical structure database systems.

---

[2] IBM Almaden Research Laboratories is reported to have a program similar to Kekulé, but no published reports exist. Eastman Research Laboratories, Rochester, NY, in a private communication, reported having some portions of the capabilities necessary to assemble a program similar to Kekulé. Fraser-Williams (Poynton, Cheshire, England) has a program for converting structure images having a constant format.

2. Description of the System

The process of interpreting a chemical structure diagram consists of roughly seven steps:

1. Scanning.
2. Vectorization
3. Searching for dashed lines and dashed wedges.
4. Character recognition.
5. Graph compilation.
6. Post processing.
7. Display and editing.

2.1 Scanning

Scanning is integrated into Kekulé but is actually accomplished using software and hardware supplied by second parties. The user of Kekulé initiates the scanning operation simply by choosing an icon in the Kekulé window with a mouse. Kekulé does provide a simple technique for finding structures interspersed with text that is based on looking for white-space surrounding any object roughly in the middle of the scanned area. This is effective for handheld scanners, but of limited value to flatbed systems. Our experience is that 300 dpi is acceptable and 400 dpi probably about optimal because the information content of characters printed on paper degrades rapidly due to optical, paper, ink wicking, and printing limitations so that higher scanning resolutions only "enlarge" the image without adding "information." TIFF and other formats can be used with Kekulé for batch processing.

2.2 Vectorization

Vectorization consists of reducing the scanned image to line elements only 1 pixel in width by thinning[3] and raster-to-vector translation. The results of this step are lists of vectors associated with the original pictorial elements by coordinates of the vector end points. One of the major problems solved in this step was that of polygon minimization or approximation. Our approach eliminates the problems of typical algorithms of creating artifact nodes and missing "real" nodes. The major changes we made in the usual algorithm [3] were in picking starting points and in adaptively setting the allowed deviation from a straight line based on the assumed length of the current segment being examined. The starting points usually selected are the beginning and end of an object (list of connected line segments). The problem with this approach that we found was that U-shaped objects, for example, could very easily end up with an artifact node in the middle of the bridge between the endpoints. Our innovation was to select not the ending node, but do a search at each stage of the algorithm for the node farthest from the current starting node and partition the object's examination at that node. This solved the problem of artifact nodes.

---

[3] Numerous sources are available. A good reference is [3].

In order to eliminate many of the artifact nodes resulting from obtuse-angled line joins, we adjusted the deviation distance allowed from a straight line based on the length between the current start and end of a potential single line segment. The adjustment was computed using the formula *dist* is 1 or *length*/10.0+.4, whichever is greater, where *length* and *dist* are measured in pixels.

2.3 Searching for dashed lines and wedges

Searching for dashed lines and wedges reduces the later processing requirements for Kekulé by finding such artifacts and converting them to single picture elements instead of unconnected vectors.

Several techniques are available for finding elements of collinear lines, including Hough transforms [4]. The theory behind the Hough transform is that points on a line, transformed from *XY* into *r*-Θ space, will result in peaks that can be distinguished from non-collinear data. In practice, we found that this approach was not satisfactory because limited numbers of pixels and the erose edges of many of the line segments. For instance, Fig. 2 illustrates that some dashed line segments consist of only a few pixels.

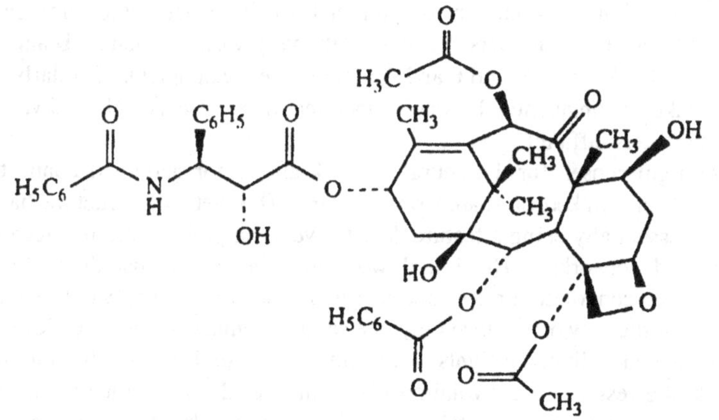

Fig. 2: Typical scanned structure (Taxol) illustrating problems encountered in interpreting: dashed lines, wedge bonds, characters ligatured to bonds and adjacent characters, and small fonts.

Another approach is to extract line segments and sort them by slope [5]. The expectation is that dashed lines will then group together because they have similar, if not identical, slopes for the individual segments. In practice, the line segments are so short that they may have widely varying slopes, or even no distinguishable slope for dashes approaching the shape of round dots.

We were, therefore, forced to search for other methods for isolating dashed lines. The method that we chose was essentially one of exhaustive search – testing of all possibilities – over only the subset of features that might be possible constituents of a dashed line or dashed wedge feature. In general, we identify all dashes that consist of at least two line segments. This includes most cases where one or more of the two line segments are attached to other bonds.

2.4 Character recognition

Character recognition (OCR) was assumed to be an easily solvable problem when we began working on this project since various techniques [6] for recognizing characters have been developed, some as long ago as 1870. Three problems with commercial approaches to OCR led us to create our own: (1) they did not work well in an environment including both graphic elements and characters; (2) recognition accuracy was very low, 80-85%, for the small fonts we found were used for most structures; and (3) they would have been relatively expensive to bundle with Kekulé at the time it was being initially developed.

Our investigation into OCR's practical problems led us to try several approaches to achieving invariance in translation, rotation, and size of characters while maintaining a high degree of accuracy. Among those investigated were Zernike moments [7,8,9], Fourier descriptors [10,11], Gabor transforms [12], coordinate transforms [13], and Hough transform-like approaches [14,15]. Introducing such invariances may reduce the dimensionality of the pure pattern recognition problem inherent in OCR, but invariance in rotation, at least, introduces other problems in distinguishing between characters that *are* rotationally variant (such as **b** and **q**; **d** and **p**; **1** and **–**; **E**, **W**, and **M**; and **6** and **9**, among a few examples). Similarly, size invariance makes distinguishing between upper and lower case as well as **5** versus **S** and other characters difficult.

The paradigm chosen for the neural network allows for patterned connections between layers and linked or shared weights [16]. This network structure has the advantage of essentially being a feature detector versus a global pattern detector for fully connected networks. The initial work to develop this paradigm did use manually created convolution maps, and that is perhaps the easiest way to envision the operation of the network. Since only one set of weights exist between layers for a pattern — shared or linked weights — and this weight set is created by training of the network, the result is a convolution map with added dimensionality since the map created during training contains floating point weights for each cell with positive and negative values. Thus, while the concept is a convolution map, the map is indecipherable. As noted, the result is a feature detector and the network exhibits vastly better generalization results than the authors ever experienced with fully connected networks.

The resulting OCR system typically achieves a raw accuracy of about 99.9% based on recognition levels of a selected test set. This set was selected based on the following criterion: was the character recognizable to the author out of context. Admittedly, this is a better test than occurs in "real life" but works better in a training/testing environment since true outliers will disturb the neural net training process. It should be remembered that our typical fonts consist of characters no larger than 7 points with subscripts as small as 3.2 points (Fig. 3 and Fig. 4). Our

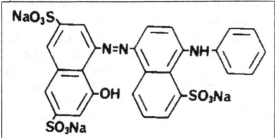

Fig. 3: Structure scanned at 400 dpi and enlarged here 200% to illustrate typical worst-case capability for the Kekulé OCR. The subscripts were about 3.2 points in the unenlarged original. The skewing was not intentional, but is typical of hand-held scanning. Kekulé interprets (and straightens) this structure perfectly.

experience in testing commercial OCR systems on small fonts produced recognition rates of about 85%. Just as commercial OCR systems utilize spell checking to achieve a higher accuracy, Kekulé combines raw data on characters with contextual (chemistry) rules to "spell check" and achieve a significantly higher effective recognition rate.

An additional technique introduced while creating the training set was to group similar characters into a single "character." For instance, in many fonts the differences between the vertical line characters (Iil1l) are indistinguishable out of context especially in sans-serif fonts. Similar cases exist for the circular characters (0oO) as well as upper- and lower-case characters (uU, vV, wW, xX, zZ, etc.) that are essentially identical in shape, varying only in size.

If one configures the neural network outputs so that each output represents a character, each output will then represent some measure of the probability of a match for the character presented to the input of the network. We took advantage of this by ranking the outputs and keeping those above an arbitrary threshold that was derived experimentally for continued processing.

Fig. 4: Portion of Fig. 3 to illustrate that 3.2 point characters at any scanning resolution are going to be difficult to interpret.

Thus, if the actual character was a poorly formed 5, the outputs might be relatively strong for both 5 and S, and we would retain both, determining which was correct from context.

Objects that are not recognized as characters are further analyzed to determine whether they are ligatured characters–characters attached to other characters or to bonds. If they are determined to be characters, the individual symbols recognized are retained for later processing along with the above-matched characters. Depending on user setup options, Kekulé may prompt for characters that are not valid (above a threshold) at this stage in processing. Alternatively, prompting can be delayed until the post-processing stage when "chemical spell checking" has indicated that characters are invalid.

Individual characters are assembled into character strings based on *XY* coordinates; that is, the *XY* positions of various individual characters are compared, and character strings — formulas, say, or an atom symbol and subscript — are assembled based primarily on adjacency of the coordinates. This is an adaptive process because strings in the original scanned documents are often separated by less space than one might expect in "normal" OCR processing. The prevalence of subscripts and superscripts and the frequency with which one encounters separate symbols that are not well separated spatially introduce additional problems, all of which are dealt with in this process.

2.5 Graph compilation

Graph compilation is the process of interpreting the remaining vector data – after eliminating those vectors associated with characters or character strings identified in the preceding step – into a connection table or graph. In the context of chemistry, this step consists of computing a connection table of bond and atom node information. Being very simplistic, if the drawing were a map, the results would be roads ("bonds"), towns ("nodes"), and the character information town names; and, similarly, for engineering diagrams. Chemical diagrams exhibit most of the general features of maps and engineering drawings, just at a very low "density."

A surprising amount of chemical intelligence was built into Kekulé to allow it to put together the graphic and character data as "chemistry" with a very high degree of accuracy as attested to by users and reviewers [17˙18˙19]. The "intelligence" consists of information about the bonding of each atom type, that a single connection between two nodes is a "single" bond, two (roughly) parallel lines between nodes a "double" bond, and so forth. In addition, Kekulé "understands" the notation chemists use for common structural groups or group formulas. The appoach taken in Kekulé was to actually parse such formulas much as is done by chemists. For instance, the formula for a carbonyl group —COOH is parsed left-to-right by noting that an external bond exists decreasing the valence of the following carbon atom by 1 (leaving 3). The first oxygen has a valence of 2 and could be bonded to the carbon and the following oxygen, except that is a rare case in chemistry and the assumption is that the entire valence should be subtracted from the carbon — leaving 1. The next oxygen thus can have only a single bond to the carbon; the final hydrogen has a single bond (valence of 1) to the previous oxygen. This is how chemists do it and essentially how it is done in Kekulé. A very limited number of special cases were defined manually to process examples that do not parse automatically.

2.6 Post processing

Rotational variance from scanning is typically only a degree or two — even this small error is visually very distressing on-screen. Scans with intentional errors of only 4 degrees appear visually to be off by a far greater amount. While rotation error is not a critical factor in OCR, it is very important in the cleaning of the result to produce a publication-quality image. Our approach to rotation correction was to

155

examine all "long" line segments, compute the angle molulo 15 (degrees), and accumulate the result into one degree "bins." The bin having the highest count and representing a rotation of less than 4 degrees was assumed to be the amount of rotation due to scanning error. The 15 degree figure was chosen because the "standard" when drawing structures is to draw at 15 degree multiples and the frequency of vertical and horizontal lines as well as hexagons is very high in all organic chemical diagrams. Another factor tending to force 15 degree multiples has been stair-casing — keeping angles to 15 degree multiples automatically tends to minimize the most distrubing aspects of stair-casing.

Some elements of interpretation are difficult to complete until the "easy" ones have been eliminated from consideration. For instance, a circle is not recognized as a "circle" but as, typically, a ring contained inside another ring. It would be feasible to find circles at an earlier stage — just as we find dashed lines, for instance — but the processing required would be considerably more complex and difficult and the heuristic we developed is "chemically" not allowed and is, hence, easily recognized.

We have not discussed bond crossings so far. They are initially recognized as nodes if there are no gaps at the crossing. Since bonds cross (in typical drawings, anyway) only when ring systems are involved, subsequent analysis of ring systems is used to determine which nodes are really bond crossings. Large brackets and parentheses, also initially recognized as nodes and bonds, are recognized by testing the configuration of their line segments.

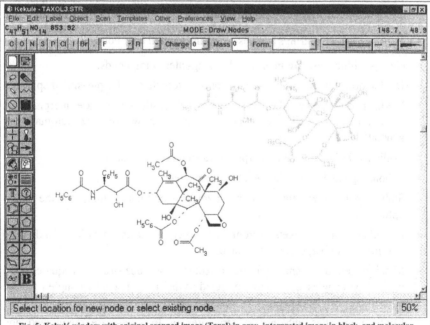

Fig. 5: Kekulé window with original scanned image (Taxol) in gray, interpreted image in black, and molecular formula and weight near the upper left corner of the window.

2.7 Display and editing

Display and editing processing consists of several operations including rotating the entire graph to adjust for scanning error (as discussed in 2.4) and cleaning up bond angles to standard values when possible. The final graph is then displayed on the video screen in black with the original, scanned image in gray (Fig. 5). With both images present simultaneously, it is easy for the user to verify whether the interpretation is correct. In addition to the interpreted graph, Kekulé computes the molecular formula and molecular weight of the interpreted graph — both useful for verifying the interpretation.

3. Interpretation Capabilities

Kekulé will correctly interpret most chemical notation. The following list represents the highlights of Kekulé's ability to read chemical structures:

- Single, double, or triple lines connecting atom symbols or intersecting other lines are interpreted as single, double, or triple bonding of atoms with carbon assumed if no atom symbol is present.

- Characters representing atom symbols, subscripts, and superscripts are found and interpreted. (IUPAC conventions are assumed: Charge to the upper right of the symbol, mass to the upper-left, and repeat counts at the lower right.)

- Group formulas are interpreted into their appropriate connection table structure while retaining the on-screen character display complete with subscripts, etc. Group formulas may be multiply bonded to the remainder of the structure.

- Vertical character strings such as $\overset{N}{\underset{H}{}}$ interpreted as if bonded.

- Circles within rings are interpreted as ring-alternating bonds.

- Double bonds within rings are positioned automatically for pleasing display.

- Wedged bonds, either filled or as a series of parallel dashes, are interpreted as stereo-up bonds. "Thick" bonds between two wedges are automatically generated.

- Collinear dashed lines are interpreted as stereo-down bonds.

- Pi bonding (such as in ferrocene) is recognized.

- Stubs can be interpreted as either hydrogen or methyl groups at the user's option.

- Parentheses and brackets surrounding structures, with specific or indefinite multipliers or charges, are recognized.

- Misalignment of scanning will be corrected to produce structures square to the page. Bond angles are further corrected to the nearest standard angle. (These adjustments can be cancelled for special cases.)

- Multiple moieties, whether as structures or formulas, are recognized and preserved.

- Bond crossings are recognized.
- Locants, node numbers used for nomenclature amplification, are interpreted and discarded.

4. Discussion and Conclusions

The current version of Kekulé was tested on 524 chemical structures obtained from a wide variety of sources. These sructures were chosen to test Kekulé's limits and are, therefore, not necessarily typical of structures[4] in general. Of those structures, 98.9% were processed with an average of .74 prompts per structure for verification of interpretation or correction. The average time for this test of Kekulé's abilities was 9 seconds per structure on an 80486 at 33 mHz. Five of the structures tested required more than 30 seconds of editing to correct interpretation errors due to pathological problems such as broken characters. Kekulé provides a solution to the long-sought goal of being able to automatically interpret chemical structure diagrams. The techniques used are generally applicable to the problems of interpreting other diagagrams such as electrical circuits, building wiring, utility maps, and road maps.

REFERENCES

[1] Rozas, R.; Fernandez; H. Automatic Processing of Graphics for Image Databases in Science. *J. Chem. Inf. Comp. Sci.* **1990**, 30, 7-12.

[2] Contreras, M. L.; Allendes, C.; Alvarez, L. T.; Rozas, R. Computational Perception and Recognition of Digitized Molecular Structures. *J. Chem. Inf. Comp. Sci.* **1990,** 30, 302-307.

[3] Pavlidis, T. *Algorithms for Graphics and Image Processing*; Computer Science Press: Rockville, MD, **1982**; pp 281-297.

[4] Rosenfeld, A.; Kak, A. C. *Digital Picture Processing*; Academic Press: Orlando, FL, **1982**; Vol. 2, pp 121-126.

[5] Kasturi, R.; Alemany, J. Information Extraction from Images of Paper-Based Maps. *IEEE Trans. Software Eng.* **1988,** 15 (5), 671-675.

[6] Govindan, V. K. Character Recognition–A Review. *Pattern Recognit.* **1990,** 23 (7), 671-683.

[7] Hu, M. K. Visual Pattern Recognition by Moment Invariants. *IRE Trans. Inf. Theory* **1962**, 2, 179-187.

[8] Khotanzad, Al; Hong, Y. H. Invariant Image Recognition by Zernike Moments. *IEEE Trans. Pattern Anal. Mach. Intell.* **1990,** 12 (5), 489-497.

[9] Teague, M. R. Image analysis via the general theory of moments. *J. Opt. Soc. Am.* **1980,** 70 (8), 920-930.

---

[4] We imposed a limit of 30 s for editing to include a structure in the prior group of 98.9%. This was an arbitrary limit and all but one of the five structures failing this test could be corrected in no more than 60 s.

[10] Persoon, E. Shape Discrimination Using Fourier Descriptors. *IEEE Trans. Syst., Man, Cybern.* **1977,** *SMC-7* (3), 170-179.

[11] Zahn, C. T.; Roskies, R. Z. *Fourier* Descriptors for Plane Closed Curves. *IEEE Trans. Comput.* **1972,** C-21 (3), 269-281.

[12] Korpel, A. Gabor: frequency, time, and memory. *Appl. Opt.* **1982,** 21 (20), 3624-3632.

[13] Reber, W. L.; Lyman, J. An Artificial Neural System Design for Rotation and Scale Invariant Pattern Recognition. *Proc IEEE 1st Int. Conf. Neural Networks* **1987,** 4, 277-283.

[14] Lashas, A.; Shurna, R.; Verikas, A.; Dosinas, A. Optical Character Recognition Based on Analogue Preprocessing and Automatic Feature Extraction. *Comput. Vision, Graphics, and Image Process.* **1985,** 32, 191-207.

[15] Kahan, S.; Pavlidis, T.; Baird, H. S. On the Recognition of Printed Characters of Any Font and Size. *IEEE Trans. Pattern Anal. Mach. Intell.* **1987,** PAMI-9 (2), 274-288.

[16] Jackel, L. D. et al; Hardware Requirements for Neural-Net Optical Character Recognition, *IEEE Int. Joint Conf. on Neural Networks,* **1990,** Vol. 2, pp 855-860.

[17] Rogers, A. Kekulé for Windows: The Complete Structure Input System. Journal of Chemical Information and Computer Science, **1994,** 34, pp 1225-6.

[18] Seiter, C. Kekulé 1.1. *MACWORLD*, November **1994,** pp 63-4.

[19] Yip., C. W., et al. Scanning for Structures. *Analytical Chemistry*, **1994,** Vol. 66, No. 24, pp 1216A-1217A.

Knowledge-Based Segmentation for Automatic Map Interpretation

Jurgen den Hartog[1]   Ton ten Kate[1]   Jan Gerbrands[2]

[1] TNO Institute of Applied Physics, P.O. Box 155, 2600 AD Delft, the Netherlands
[2] Delft University of Technology, Fac. of Electrical Engineering, P.O. Box 5031, 2600 GA Delft, the Netherlands

Abstract. In this paper, a knowledge-based framework for the top-down interpretation and segmentation of maps is presented. The interpretation is based on a priori knowledge about map objects, their mutual spatial relationships and potential segmentation problems. To reduce computational costs, a global segmentation is used when possible, but an applicable top-down segmentation strategy is chosen when errors in the global segmentation are detected. The interpretation system has been tested on utility maps and the experiments show that when a top-down resegmentation strategy is used to correct errors in the global segmentation, the recognition performance is improved significantly.

1 Introduction

Our research concentrates on the automatic conversion of Dutch utility maps and in this paper we consider the problem of obtaining a correct segmentation of the grey value map images. Over the past eight years a number of interpretation systems for line drawings have been described in literature. In most of the reported work, the segmentation process is assumed to be trivial as a binary scanning process is employed [21], or simple techniques such as thresholding are applied [19, 22]. Furthermore, many of the interpretation systems have no strategy to handle segmentation errors and therefore assume good quality binary images to be processed [1, 15, 26]. In some work, the problem of an imperfect preprocessing result is recognized. These systems approximate the graphics by using vectorization and try to solve local errors which might be due to the segmentation, such as broken lines, by using techniques based on merging collinear vectors separated by a small gap, e.g. [14, 18]. Though such an approach might be successful in more or less simple situations with carefully drawn high quality maps, the result might be less satisfactory for more complex applications. For example, due to years of intensive use the maps may be wrinkled and stained, while often parts have been erased and redrawn, leaving traces of ink on the linen material.

The most commonly used technique of thresholding introduces the problem of determining the optimal global threshold [20, 23]. If the threshold value is too low, small objects are lost, while a high threshold value results in noisy images and smearing of the objects. Often, there is no optimal threshold level which avoids smear and loss of objects. In such a case, a practical solution is to determine a sub-optimal threshold which minimizes both effects. Better results may be obtained by using local adaptive threshold techniques, e.g. [3, 8], but these techniques suffer from a variety of disadvantages.

For example, usually a threshold is calculated from a window of interest. If the size is too small, these algorithms tend to emphasize noise or paper texture in regions without foreground pixels. In the case of larger windows, the algorithm becomes computationally expensive and it may have the same drawbacks as global threshold techniques.

An interesting approach to top-down interpretation and segmentation of drawing images is Anon, a schema-based system described by Joseph and Pridmore [13]. This system comprises a set of schemas which represent the entities to be recognized in the image. Each schema consists, among others, of a geometrical object description and a number of C-functions to interface to the image processing stages. The geometrical description has to be satisfied by the results of image processing before the instantiation of a schema class. This paper is one of the first to observe that there is a need for top-down segmentation strategies in drawing conversion. In Anon, all objects, without exception, are extracted by means of top-down segmentation. A typical A0-sized drawing (1 square meter or circa 1600 square inches) contains thousands of objects and if all these objects have to be extracted by means of top-down segmentation the system may become very slow and thus obstruct a practical solution.

Even if the initial global segmentation contains many errors, it still contains useful information which should not be discarded. Hence, a more realistic approach might be to develop a strategy which combines both bottom-up and top-down segmentation techniques. The computationally cheap global segmentation is used when possible, but specialized top-down segmentation techniques are utilized when needed.

In this paper, a knowledge-based strategy is proposed which combines low-level bottom-up processing with top-down segmentation. The low-level preprocessing is discussed first. The top-down segmentation is embedded in a framework for contextual reasoning which is described in Sec. 3. In Sec. 4 the ideas underlying top-down segmentation are introduced, while the organization and representation of the knowledge is discussed in Sec. 5. The top-down segmentation is illustrated with two examples in Sec. 6. The experiments and the results are then given in Sec. 7. The advantages and limitations of our approach are discussed in Sec. 8 and in Sec. 9 our conclusions are given.

2 Preprocessing

Although the top-down segmentation strategy can correct errors made during bottom-up segmentation, the initial preprocessing is still important to the final interpretation result. The preprocessing to obtain a good low-level description of the image contents consists of the following steps:

1. Sharpening to enhance the blurred original image.
2. Binarization.
3. Removal of holes in the graphics.
4. Decomposition of the graphics into primitives.

2.1 Image Sharpening

The use of standard linear filters such as the Laplacian operator to sharpen the blurred image (e.g. [10]) has the advantage of computational efficiency. Disadvantages of these

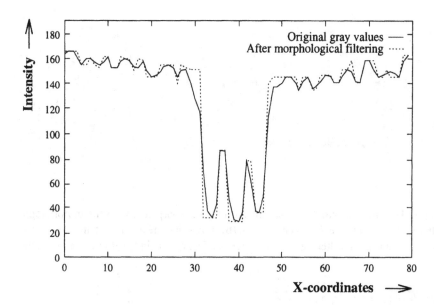

Fig. 1. The result of morphological filtering in the one-dimensional case is shown. The continuous line represents an 80 pixels wide horizontal scanline. The result after filtering is represented by the dashed line.

algorithms are the tendency to amplify noise and the necessity of clipping or scaling to make the resultant pixels span the range $< 0, 255 >$. In this section, we give a brief description of an algorithm which uses standard grey value morphology which does not have these disadvantages.

Grey value morphology is based on grey value dilations and erosions. In the case of a grey value erosion, for example, the image is scanned with a moving structuring element. The minimum value within the structuring element is calculated for each image position. This local minimum value is then stored in the pixel that corresponds with the center of the structuring element. For an extensive description of mathematical morphology the reader is referred to [24].

In the algorithm which was proposed first by Kramer [16], for each pixel the local grey value minimum and maximum within a structuring element is computed. The algorithm simply consists of replacing the grey value of each pixel by its local minimum or maximum, whichever of the two is closer in value. The result of the filter in the one-dimensional case is shown in Fig. 1.

Kramer showed that repeated filtering always converges, although usually many iterations are needed for complete convergence. However, after a small number of iterations, only few pixel values will change. For the utility maps only 3 iterations with a 3x3 window are sufficient for near-complete convergence. Further, it was found that filtering with a 3x3 sized window and a small number of iterations yielded better results than filtering once with a larger window. This is illustrated in Fig. 2. In Fig. 2a, a blurred dimension is shown. Applying the filter once with a 7x7 window results in a poor con-

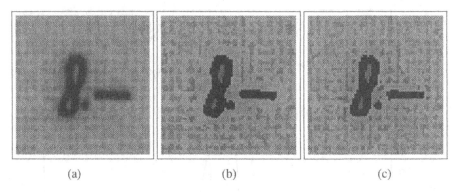

(a) (b) (c)

Fig. 2. The small original 64^2 map image in (a) is sharpened with the morphological filter with window size 7x7. The result in (b) shows that the main digit and the dot have become connected. Filtering with window size 3x3 results in (c) after three iterations.

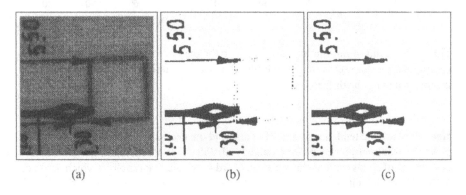

(a) (b) (c)

Fig. 3. The original 200^2 map image in (a) is thresholded at a level for the optimal extraction of the graphics. The binary image in (b) shows that this may result in a noisy image. However, if the image is thresholded with hysteresis this results in the much less noisier image shown in (c).

trast between the main digit and the dot (Fig. 2b). The result with a 3x3 window after 3 iterations is shown in Fig. 2c.

2.2 Thresholding with Hysteresis

In the introduction, the advantages and disadvantages of local algorithms [3, 8] and global threshold algorithms [20, 23] were discussed. Another approach to obtain an initial segmentation is *thresholding with hysteresis* [4, 25]. In this method, a segmentation is obtained in two steps. In the first step, all pixels are classified as one of three possible categories by using a high and a low threshold. If a pixel has a grey value below the low threshold, it is classified as a definite object pixel while pixels with grey values above

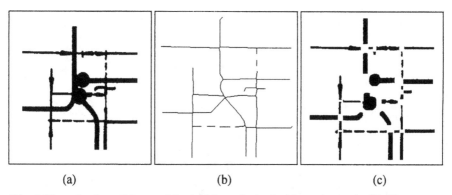

| (a) | (b) | (c) |

Fig. 4. Vectorization of the graphics in (a) results in the binary image in (b). The reconstructed primitives are displayed in *exploded view* in (c).

the high threshold are classified as definite background. The pixels with values between the high and the low threshold remain to be processed further in the second step.

Though the remaining pixels usually are object pixels, they frequently correspond to noise or stains in the map. Only if they are connected to any of the definite object pixels are they considered to be object pixels too. Because the previous image sharpening step reduces the number of actual grey levels in the image, the thresholding step will be less sensitive to the selection of the threshold values. Fig. 3 shows an example of global thresholding and thresholding with hysteresis.

2.3 Hole Removal

After thresholding the grey value image, the binary image is processed to remove small holes. First, based on the area, for each object a distinction is made between small objects such as characters and large objects such as the graphics. Holes are only likely to occur in small objects, e.g. digits such as a 6 or a 9. Therefore, small holes in the graphics can be regarded as noise and are removed.

2.4 Decomposition of the Graphics into Primitives

Skeleton vectorization is an often-encountered step in the interpretation of maps, engineering drawings and other line drawings [1, 2, 9, 11, 12, 15]. However, in spite of the attractive simplicity, rigorous vectorization may introduce unwanted inaccuracies. In Fig. 4a, a piece of graphics from a utility map is shown with its vectorization in Fig. 4b. Reducing the graphical objects to vectors discards the morphological information required to recognize the various objects. In [7] we described an algorithm which decomposes a binary image into its constructing primitives. In the remainder of this paper, we refer to a primitive as the most basic image component that consists of a set of connected pixels without any meaning attached to it yet. Fig. 4c shows the resulting primitives. To be able to display individual primitives, the primitives are shown in the *exploded view*.

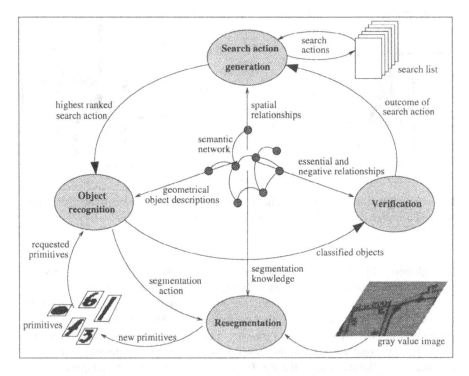

Fig. 5. The concept of contextual reasoning and resegmentation

Describing graphics in terms of primitives has the advantage that no morphological information is lost, as opposed to a vector description. Another advantage is that all primitives can be stored in separate bitmaps, which enables the image processing of individual primitives without interference from the neighborhood in the original image. As a consequence, it is possible to calculate many useful features for individual primitives, which is not possible with a vector-based description. The primitive-based description clearly facilitates object recognition when compared to a vector description and it forms the basis for the remaining interpretation process.

3 Contextual Reasoning

The strategy to combine both bottom-up and top-down image processing is implemented in a framework provided by a knowledge-based map interpretation system described in [6]. This system guides the interpretation by means of contextual reasoning.

The concept of contextual reasoning is based on alternating a top-down process to generate search actions, a bottom-up process to recognize objects and a process to verify the results. The interpretation cycle consisting of these three processes is shown as part of a larger process in Fig. 5. The bottom-up process receives from the top-down process a search action, i.e. a task to search for a specified object type in a restricted area. All primitives overlapping with the search area are matched with an a priori geometric specification consisting of a list of features and for each feature a range of allowed values. If the primitive corresponds with the geometrical specification, it is offered as a candidate for classification to the verification process. If no inconsistencies with former results are detected, the search results in an identification and new search actions are added to the search list.

The mechanism to generate new search actions is based on the perception that each object type is related to other object types. For example, in many applications a distance between two objects is depicted by an arrow and a dimension, a numerical representation of the depicted distance. Thus, detection of an object immediately generates expectations about other objects in its neighborhood, which are very suitable to generate new goals for the interpretation process.

For example, in the case of detection of an arrow, search tasks for related objects, such as the dimension, are generated for a small region of interest (ROI) around the arrow. Each time a new object is detected, the top-down process collects all related object types and generates new search tasks for the search list. The search task at the top of the search list is then distributed to the bottom-up process which tries to detect new objects in the assigned ROI. Because a search for an object is based on contextual evidence and the search area is restricted to a small and confined part of the image, both the number of search actions and the number of incorrect object classifications are reduced, thus rendering the interpretation both efficient and reliable.

4 Top-down Segmentation

4.1 Inconsistency-based Resegmentation

The concept of top-down segmentation is integrated within the contextual reasoning framework. During contextual reasoning, inconsistencies can be detected between recognized objects and prior knowledge. In our application area we experienced that inconsistencies are often due to a poor initial segmentation. If knowledge about inconsistencies and potential segmentation problems is used during interpretation to improve the local segmentation, many inconsistencies may be solved automatically.

Inconsistency Detection The utility maps considered in our research are 10 to 30 years old and have been used in field work. As a consequence, the maps are often wrinkled and stained and it is more than likely that, due to noise or poor segmentation, many objects are misclassified or rejected from classification. These misclassifications and rejects slow down the interpretation process considerably as the results from the automatic interpretation have to be verified manually. It is therefore important to be able to detect and solve misclassifications and rejects during the interpretation instead of afterwards.

Because the aim of the research is to develop a system with a broad application area, our research has concentrated on a method based on the detection of contextual inconsistencies between results and prior knowledge. The method is based on the perception that three types of spatial relationships between objects may be distinguished:

1. The *optional* relationship.
2. The *essential* relationship.
3. The *negative* relationship.

An optional relationship between two objects indicates that if the first object is found, it is likely that the second will be found in its vicinity. For our application, most of the relationships are optional, but a substantial part of the relationships is essential. An essential relationship implies that if the first object is found, the related object *must* be present too. An example of an essential relationship is the arrow which always has a dimension. The *negative* relationship is the opposite of the essential relationship. In this case two objects should not share a specific spatial relationship. For example, the distance between a conduit and a house should not be less than 0.5 meters.

Inconsistency Handling The concept of inconsistency handling is based on knowledge about the causes of specific inconsistencies. For example, each house has a number. Thus, if a house has been found but the number cannot be detected, in general there are three possible causes for this inconsistency:

1. The classification of the house is wrong.
2. The draftsman did not write the number on the map.
3. The house number cannot be recognized due to poor segmentation.

Knowledge about these causes can help to solve the inconsistency without the assistance of the operator. To check whether the classification of the house is correct an alternative detection method is applied first. If the second technique confirms the classification of the house it is assumed that the segmentation of the house number is poor. In this case, the local segmentation could be improved by using specific algorithms with parameter settings tailored to an optimal segmentation of small objects such as digits. If resegmentation does not result in the recognition of the house number, this might be due to a violation of the drawing rules and the system should invoke operator assistance.

The interpretation cycle is shown in Fig. 5. Each filled ellipse represents a sub-process within the interpretation process. If, for a given search action, the outcome of object recognition is such that the verification process detects an inconsistency, a message is passed to the search action generation process. This process then searches in the knowledge base for a segmentation method to solve the inconsistency. If a method is available, it is passed to the object recognition process which sends the segmentation action to the resegmentation process. Next, the resegmentation process gathers the segmentation algorithm and the corresponding parameter settings from the semantic network and, as a next step, executes the segmentation action. After resegmentation, the resulting primitives are added to the existing list of primitives and the object recognition process then again executes the search action which led to the inconsistency. If the resegmentation was successful the inconsistency will no longer be detected. If the inconsistency is not solved, the classification causing the inconsistency is rejected and it remains to be classified by the operator afterwards.

4.2 Directly Processing the Grey Value Image

If it is known a priori that the initial segmentation cannot properly segment certain object types, it is inefficient to use the inconsistency mechanism for resegmentation. A better approach is to process the grey value image directly to find the objects. The interpretation process is therefore extended with the addition of grey value objects. When searching a grey value object, a segmentation method and a geometrical object description are passed to the object recognition process. Object recognition then passes the segmentation action to the resegmentation process and after resegmentation it tries to match the segmentation result with the object description.

5 Knowledge Representation

Flexible manipulation of the knowledge is important when developing an interpretation system. For this purpose, an explicit knowledge representation language has been developed. In this section, the merits of explicit knowledge are discussed first and are followed by some relevant aspects of the language.

5.1 Explicit Knowledge

Public utilities usually provide multiple services such as electricity, water, gas and cable television, and for each service a different type of map is used. Besides the multiple services, multiple networks are used for the transportation of a single service. For example, gas is transported to local distribution centers through a high pressure network, while for distribution to customers a low pressure network is used. All these types of networks are drawn on different maps. The possible variety in maps becomes wider when the drawing conventions change in time or when services are taken over from other public utilities.

Because many applications have to be considered even for a single public utility, flexibility of interpretation is one of the most important goals in our research. Although the symbols and structure of the maps are more or less similar for the various applications, it may be clear that an interpretation scheme for one type of map cannot be used directly for other maps. However, the concepts underlying the interpretation are shared by all map applications. If a priori knowledge about an application is separated from the implementation, it should be relatively easy to tailor an interpretation system to other applications.

In our system, the knowledge has been separated from the implementation by means of a knowledge specification language. All a priori knowledge about the application is read from a file at run time and converted into an internal data structure. This data structure is a hybrid structure which is referred to in this paper as the semantic network. All knowledge can be adapted at run time, and, as a consequence, the time needed to develop an interpretation system is reduced significantly as reprogramming can be circumvented.

5.2 Basic Objects and Relationships

The concept of contextual reasoning is based on the assumption that all objects in a map are interrelated. Basically, the procedural knowledge used by the contextual reasoning mechanism should specify when to search and where to search for an object, while the declarative knowledge should describe how to recognize objects. The procedural knowledge is represented by the spatial relationships between objects. The declarative knowledge is represented as a feature-based geometry description for each object type.

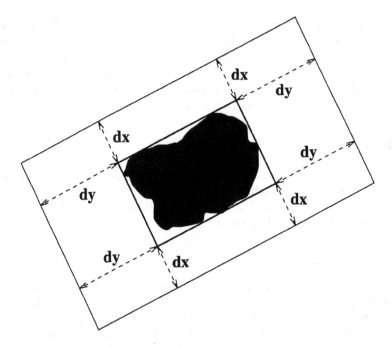

Fig. 6. The search area is based on the extension of the MAER of an object.

Object Descriptions Object recognition is based on matching each applicable primitive, i.e. each primitive overlapping with the search area, with an allowed range of attribute values in the object description. The description offers a large set of standard features such as area, length, width along the skeleton and a set of features based on the minimum area enclosing rectangle (MAER). For a detailed overview of all attributes the reader is referred to [6].

The syntax of the object description is simple as is demonstrated in the following example describing the object type *double headed arrow*:

```
DEFINE OBJECT
    objname     DoubleArrow
    id          10
    max_width   [4.0, 7.5]
    avg_width   [1.4, 7.0]
    length      [60.0, 1000.0]
    maer_ratio  [0.0, 0.4]
    maer_width  [0.0, 20.0]
    my_func     [-17.1, 22.5]
ENDDEF
```

Besides the standard set of attributes, it is easy to expand the object description by the addition of new functions. This is illustrated in the example above where a new function is added named *my_func*.

Relationship Descriptions The description of the relationships between objects (Sec. 4.1) is similar to the object description. The standard attributes of the relationship description include a unique id, the object names of the two object types involved, the type of the relationship, a priority number and a specification of the search area. Analogous to the object description, a set of standard relational features is provided, such as inclusion and angle. Again, the user can extend the set of features by adding new functions.

```
DEFINE RELATION
        id              100
        from            DoubleArrow
        to              Dimensioning
        type            Essential
        priority        3
        search_area     <70,0>
        my_rel_func     [2.1, 2.3]
ENDDEF
```

The example indicates that if a double-headed arrow is found, it is likely to find a dimension in its vicinity. Based on this relationship, a search action for the dimension is generated and added to a search action list which is sorted on the priorities of each search action. In this case, the search action is assigned priority 3. The search area is based on the MAER of the previously detected object. The search area has two arguments, dx and dy, specifying the extension of the width and the length of the search area respectively. Fig. 6 shows an example.

5.3 Knowledge for Resegmentation

The knowledge needed for resegmentation of the grey value image is represented in a way similar to the previous examples. The segmentation knowledge has to be tailored to optimally extract a single object which cannot be found by using the initial segmentation. Since the resegmentation will be carried out locally, the knowledge description has to provide an argument to specify the size and the location of the part of the image to be segmented. It should also be possible to specify the image processing functions, their arguments and the execution order of the functions.

```
DEFINE GRAY_OBJECT
        id              20
        from            DoubleArrow
        to              House
        reseg_area      <10,100>

        median_filter   5 5
        local_thresh    31 31 0.4

        max_width       [3.0, 5.0]
        avg_width       [2.6, 3.4]
        my_house_func   [1.0, 4.0]
ENDDEF
```

In the example above, the knowledge specification for the optimal segmentation *and* recognition of a grey value object is shown. The segmentation description is applicable to the situation when a double-headed arrow is found and a house segment is searched directly in line with the arrow. The part of the image to be resegmented is calculated from the MAER of the arrow and the reseg-area specification. Each image processing function that is defined in the knowledge specification, is succeeded by an argument list. In the example, two image processing techniques are applied to the grey value image. First, the image is processed with a median filter to remove the paper texture. Next, the filtered image is segmented by using a local and specialized thresholding algorithm. The binary image that results contains new primitives which are added to the list of existing primitives. If the resegmentation was successful, a search action for the house using the description in the remainder of the object specification should result in its recognition.

6 Two Examples

The application for our research is introduced first and is followed by two examples to illustrate the top-down resegmentation process.

6.1 The Application

The main concern of the public utility is to convert the position of the pipelines and conduits on their maps to a digital description in world coordinates acceptable by a GIS. Currently, we consider the conversion of the low-voltage electricity maps. On these maps, the relative position of the conduit is depicted by the perpendicular distance between the conduit and clearly distinctive landmarks such as the corners of the houses. On the utility maps, the distance between two points is represented by an arrow and a dimension. A dimension consists of multiple objects which can be categorized as either a dot or a digit. Since the exact location of the houses can be obtained in a digital format from the Dutch cadaster, the recognition of the conduit, houses, arrows and their dimensions should be sufficient to reconstruct the exact position of the conduit.

Unfortunately, this rather simple model has to be extended. Often, objects such as houses and roads are drawn on the back of the map, while some arrows are drawn partially on the front and partially on the back. Fig. 7 shows an example where the head of the arrow is drawn on the front while its tail consists of the outline of a road drawn on the back. Fig. 3 shows the even more difficult case where the head of the arrow is drawn on the front while its tail is also part of a house drawn on the back. The main reason for drawing objects on both sides of the paper is efficiency when updating the maps. The electricity infrastructure is not static and both the positions and contents of the conduits are often subject to change. It is, therefore, necessary to update the paper maps regularly. For the draftsman, redrawing the situation on a map is much more convenient if the most static part of the objects on the maps, i.e. roads and real estate, are drawn on the other side.

6.2 Example #1, Segmentation of the Dimension

The dimension is a very important object type to recognize properly as it represents the numeric value of the distance indicated by an arrow. The position of the conduit is indi-

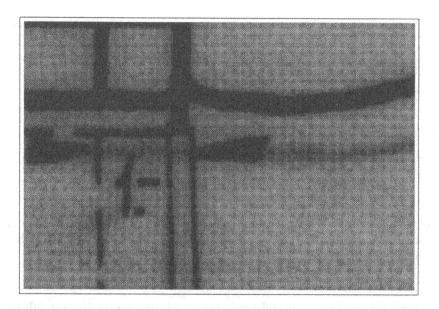

Fig. 7. An arrow with its head drawn on the front and the tail drawn on the back.

cated on the map by an arrow depicting its distance to the houses. The proper segmentation of the arrow and its dimension is thus vital for the success of the interpretation. However, the global segmentation is optimized for the graphics in the image and as a consequence the segmentation of the dimensions is often less adequate.

A dimension is made up of either three or four digits and a dot, or, a digit, a dash and a dot. For the reliable recognition of the dimension all these symbols have to be segmented correctly, but due to the global segmentation, small objects may be lost or separate objects may become connected. Fig. 8a shows the grey value image of a conduit segment and, perpendicular to it, an arrow with its dimension. Fig. 8b shows that the initial segmentation results in an acceptable segmentation of the arrow while the segmentation of the dimension results in the loss of small objects.

After recognition of the conduit, a search action is generated for the double-headed arrow. When the arrow is detected, the proper recognition of the dimension will fail due to its poor segmentation. In this situation where a double-headed arrow is drawn perpendicular to the conduit, a specific drawing rule requires the dimension to be drawn approximately midway between the arrow heads on either the left-hand or the right-hand side of the arrow. Therefore, a very specific resegmentation action for the dimension can be generated for a small neighborhood. In this example, the blurred original image is sharpened only once and then thresholded at an appropriate threshold level. The result of local resegmentation is shown in Fig. 8c.

6.3 Example #2, Segmentation of the Houses

The second example is a more complex segmentation problem. In Fig. 9a, an arrow and its dimension are shown together with the fragment of the front of a house. The house

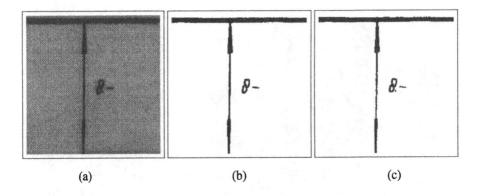

(a)	(b)	(c)

Fig. 8. The grey value image of an arrow and its dimension is shown in (a). The global segmentation in (b) shows a poor segmentation of the dimension. A local specialized segmentation results in (c).

fragment is drawn on the back of the map. There are two causes for the more difficult segmentation of the graphics on the back. The maps are often made of linen and there is interference between the texture of the linen and the objects drawn on the back of the map. Compared to the objects on the front, the range of possible grey values is much wider for the objects on the back. A simple and cheap global threshold operation would result in either a very poor segmentation of the houses or a very noisy image. Thresholding with hysteresis, which is used for the initial segmentation, would discard the houses entirely. The result of thresholding the image with a global threshold value is shown in Fig. 9b.

Similar to the first example, knowledge about the context of the houses can be used to generate top-down segmentation actions. For example, if the conduit, the double-headed arrow and its dimension are found, the region of where to expect the house is limited to a small strip directly in line with the arrow and a strip perpendicular to the arrow. Therefore, two object specific resegmentation actions can be generated for limited parts of the image. From the detected house segments new segmentation actions can be generated to detect other house segments. In this case, the resegmentation procedure consists of two relatively expensive operations. To prevent distortion by the paper texture, a standard median filter is applied. After filtering the texture, a local threshold operation [3] is applied which leads to the result shown in Fig. 9c.

7 Experiments and Results

7.1 Data Acquisition

Two original 30-year-old linen maps were available for our experiments. These A0-maps are hand-drawn to a scale of 1 to 500. Each map was scanned in 256 grey values at a density of 400 dpi. The first map was used to optimize the interpretation system while

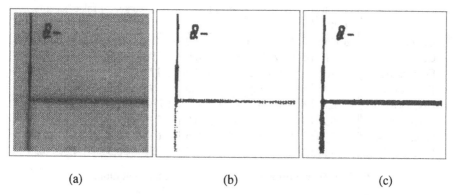

         (a)   (b)   (c)

Fig. 9. The grey value image in (a) shows an arrow, its dimension and, drawn on the back, the outline of a house. Global segmentation results in (b) while in (c) the result after a local specialized segmentation is shown.

the second map was used for testing and evaluation. There was no overlap between the maps.

7.2 Evaluation of the Experiments

To be able to compare the results of top-down segmentation with straightforward bottom-up segmentation, it is important to develop an evaluation strategy for the segmentation. In literature, several approaches to segmentation evaluation have been proposed, such as a uniformity criterion for regions [17], visual criteria for map images [27] and a criterion based on the accuracy of measurements compared to measurements from a reference image [28]. In this paper, however, we consider a specific application and the aim is to improve the recognition performance. To evaluate the top-down segmentation of the utility maps we therefore propose an evaluation criterion based on the recognition performance on the segmented objects. The main assumption in this approach is that an improvement of the segmentation stage will lead to an increase in the number of correctly classified objects. An important advantage of a recognition-based evaluation is the possibility to quantify the segmentation performance. Moreover, if the classification of each object in the test set is made available once as part of the ground truth, the experimental results can be evaluated automatically.

For each object class, the evaluation is limited to the following cases:

1. *correct*, the percentage of correctly classified objects with respect to the total number of objects in that specific class.
2. *reject*, the percentage of the objects which remained unclassified with respect to the total number of objects in that specific class.
3. *misclassification*, the proportion of all the incorrectly classified objects with respect to the total number of objects in that specific class.
4. *false accepts*, the proportion of objects from other classes which were accepted incorrectly with respect to the total number of accepts for this specific class. Note that

	ground truth	correct		misclass.		rejects		false accepts	
		#	%	#	%	#	%	#	%
Conduit *m*	8.10m	7.90m	97.6	0	0.0	0.20m	2.4	0.00m	0.0
Conduit #	712	620	87.1	0	0.0	92	12.9	1	0.2
Front arrows #	463	339	73.2	0	0.0	124	26.8	0	0.0
Back arrows #	96	0	0.0	0	0.0	96	100.0	0	0.0
Houses *m*	7.32m	4.50m	61.5	0.05m	0.7	2.77m	37.8	0.02m	0.5
Houses #	535	315	58.9	0	0.0	220	41.1	2	0.6
Dim. digit #	1310	649	49.5	6	0.5	655	50.0	13	2.0
Dim. dot #	440	216	49.0	2	0.5	222	50.5	12	5.3

Table 1. Recognition performance *with* resegmentation.

in general a false accept may be counted as a misclassification of another object class.

To calculate the above classification statistics automatically, the list of primitives is compared with the ground truth. For each vector and for each symbol in the ground truth the overlapping primitive is searched first. There are four options:

1. No overlapping primitive could be found and the number of rejects is increased by one.
2. The overlapping primitive remained unclassified and again the number of rejects is increased by one.
3. The overlapping primitive was wrongly classified and the number of misclassifications is increased by one.
4. The overlapping primitive was classified correctly and the number of correct classifications is increased by one.

Following this step, the primitives which were labeled as objects of a specific class, but could not be matched with the ground truth, can then be classified as false accepts.

7.3 Results

The interpretation is based on the classification of the primitives which are extracted either from the binary image (Sec. 2.4) or during top-down resegmentation. In general, arrows, as well as the dots and digits are represented by a single primitive. However, houses and conduits are usually represented by several primitives, and as a consequence, it may happen that these objects are only partially recognized. A partial recognition can be evaluated in two ways: by the *number* of recognized primitives with respect to the *total number* of primitives which make up the object, or by the *length* of the recognized primitives with respect to the *total length* of the primitives composing the object. The former method of evaluation is related to the number of interactions to correct the result, whereas the latter method is related to the conversion throughput. Both evaluation methods are relevant measures of the performance of the system, depending on whether the interest is focused on the number of corrective actions or on the length to be reclassified.

From a practical point of view, it seems sensible to provide both results for the object types house and conduit. In the result tables the performance in meters is denoted with an *m* while the performance in the number of primitives is marked with a #.

In Table 1, the recognition performance with contextual reasoning and top-down segmentation is presented. Besides evaluation of the overall recognition performance, it is equally important to evaluate the effect of the top-down resegmentation. Therefore, the recognition performance has to be compared with an interpretation strategy without top-down resegmentation. Table 2 shows the recognition performance on the same data that was used in the first experiment, however, without using knowledge about top-down resegmentation. In this case, it was not possible to detect any of the objects drawn on the back of the map or to solve inconsistencies automatically.

Since the conduit is the thickest entity in the image and is always drawn on the front of the map, it is the easiest object to segment and to recognize and therefore no top-down segmentation strategy was needed. As can be expected, the recognition performance did not change without resegmentation and remained at 97% and 87% respectively without any misclassifications and almost no false accepts. The difference in performance on the conduit measured in length and number of primitives can be explained by the high rejection rate of very short conduit segments.

Most of the arrows are drawn on the front but about 20% are drawn partially on the front and partially on the back. Because of their different appearance, we decided to distinguish between "front arrows" and "back arrows" in the result tables. None of the back arrows were detected, and their detection is still one of the major problems which remains to be solved. Their proper recognition is obstructed by the difficulty in recognizing the small-sized arrow heads reliably. These arrows often lack a clear context from which they can be detected. If the arrow heads cannot be recognized, it is very difficult to generate an accurate and successful segmentation action for the tail.

Even when a distinction is made between the front arrows and the back arrows, still 27% of the front arrows remain unclassified. Several causes can be identified for these rejects. The main cause is the contextual reasoning mechanism; an arrow is only searched if the related object was detected earlier. Thus, if the related object was not found, the arrow cannot be recognized either. Another cause of rejects is the decomposition process. Arrows are often intersected by lines and, as a consequence, such an arrow is decomposed into multiple primitives, which may obstruct its recognition.

In the maps used for the experiment, all houses are drawn on the back and can therefore only be extracted with resegmentation. Approximately 60% of the houses are detected correctly and only a very small number are misclassified, while the rate of false accepts is small (0.5% and 0.6% resp.). The reject rate of approximately 40% can be explained entirely by the contextual reasoning mechanism. Since the interpretation fails to classify a significant number of arrows, no segmentation actions were generated to detect the related houses. Furthermore, about 15% of the houses on the map are not related to any other objects and cannot be found with the contextual reasoning approach.

About half of the dimension digits and dots are correctly classified while very few are misclassified. Again, the rather high rejection rate can be explained almost entirely by the contextual reasoning mechanism. The dimensions are only searched if the corresponding arrow is found, but all of the arrows drawn on the back and 27% of the front

	ground	correct		misclass.		rejects		false accepts	
	truth	#	%	#	%	#	%	#	%
Conduit *m*	8.10m	7.90m	97.6	0	0.0	0.20m	2.4	0.00m	0.0
Conduit #	712	620	87.1	0	0.0	92	12.9	1	0.2
Front arrows #	463	339	73.2	0	0.0	124	26.8	0	0.0
Back arrows #	96	0	0.0	0	0.0	96	100.0	0	0.0
Houses *m*	7.32m	0	0.0	0	0.0	7.32m	100.0	0.00m	0.0
Houses #	535	0	0.0	0	0.0	535	100.0	0.00m	0.0
Dim. digit #	1310	392	29.9	6	0.5	912	69.6	5	1.3
Dim. dot #	440	125	28.4	1	0.2	314	71.4	8	6.0

Table 2. Recognition performance *without* resegmentation.

arrows remain unclassified. The false accept rate of the dimension dots is caused mainly by transposition of the dot and dash in a dimension. If a dimension represents a whole number, it consists of a main digit, a dot and a dash. Sometimes these dashes are very small and easily mistaken for a dot. The misclassified digits in Table 1 are all dashes classified as a dot, thus causing 50% of the falsely accepted dots. The importance of top-down resegmentation for the dimensions is clearly illustrated by Table 2. If only the initial segmentation is available, the recognition performance drops to 29%.

The computational costs are acceptable. The processing of an A0-sized map, scanned in grey value at 400 dpi (16400×14000), on a Sun Sparc 20, including preprocessing, interpretation and resegmentation requires approximately 45 minutes.

8 Discussion

The proposed generic framework allows for the representation of knowledge for the detection and handling of segmentation problems. Multiple specialized segmentation algorithms can be used during interpretation where each algorithm is applied only when necessary. Since limitations of the initial segmentation can be corrected during the interpretation, it is possible to obtain a significant increase in the final recognition performance. The resegmentation framework, which is driven by generic events such as inconsistencies and object detections, should be applicable to other application areas. Its use for the recognition of roads in aerial images has been demonstrated in a case study described in [5].

The explicit representation format allows the design of a resegmentation strategy at run time. This concept increases the system's flexibility and yields a significant reduction in the time and effort required to adapt the knowledge to a specific application.

The current concept has some limitations. The parameter specification of the image processing functions in the knowledge file can only handle static predefined parameters. The flexibility could be improved if during the interpretation the parameter values can be adapted to local variations in the map.

In the current concept of top-down segmentation, the generation of resegmentation actions strongly depends on contextual evidence. If the context for these actions becomes more complex, it is no longer possible to describe the context in the current knowledge base. This limitation is clearly demonstrated by the inability to detect any of the "back arrows".

9 Conclusions

A new framework has been presented which guides the interpretation and segmentation by using a priori knowledge. Because the framework allows the combination of multiple segmentation algorithms, each specialized in the segmentation of a specific object, the local segmentation of objects can be improved when needed. In the experiments, a significant increase in the recognition performance was obtained by using top-down segmentation.

The results indicate that a fully automatic system is not yet feasible; however, the developed techniques can be very useful to assist the operator in a semi-automatic environment. During semi-automatic conversion, the operator selects parts of the image which appear to be suitable for automatic conversion. The results of the automatic interpretation are displayed and can then be accepted, rejected or manually adjusted by the operator. In the latter case, the system can assist again by guiding the operator to inconsistent or rejected parts of the image. If the operator solves an inconsistent situation or reject, the automatic interpretation may continue again.

From the experiments, we conclude that the current model for knowledge representation is too limited to handle very complex situations. In the model, all procedural knowledge is represented by spatial relationships. As the model becomes more and more complex, and the number of spatial relationships and potential inconsistencies increases, the need to further structure the knowledge base will arise. Our future research, therefore, has to concentrate on the design of a more sophisticated knowledge representation model.

References

1. D. Antoine. CIPLAN, a model-based system with original features for understanding French plats. In *First Int. Conf. on Document Analysis and Recognition (Saint Malo)*, volume 2, pages 647–655, October 1991.
2. D. Benjamin, P. Forgues, E. Gulko, J.B. Massicotte, and C. Meubus. The use of high-level knowledge for enhanced entry of engineering drawings. In *9th IAPR Int. Conf. on Pattern Recognition (Rome)*, volume 1, pages 119–124, November 1988.
3. J. Bernsen. Dynamic thresholding of grey-level images. In *8th IAPR Int. Conf. on Pattern Recognition (Paris)*, pages 1251–1255, 1986.
4. J. Canny. A computational approach to edge detection. *IEEE Trans. Pattern Analysis and Machine Intelligence*, 8(6):679–698, 1986.
5. M.E. de Gunst and J.E. den Hartog. Knowledge-based updating of maps by interpretation of aerial images. In *12th IAPR Int. Conf. on Pattern Recognition (Jerusalem)*, volume 1, pages 811–814, October 1994.
6. J.E. den Hartog, T.K. ten Kate, and J.J. Gerbrands. Knowledge-based interpretation of public utility maps. accepted for publication in *Computer Vision and Image Understanding*, 1995.
7. J.E. den Hartog, T.K. ten Kate, J.J. Gerbrands, and G. van Antwerpen. An alternative to vectorization: decomposition of graphics into primitives. In *Third Annual Symposium on Document Analysis and Information Retrieval (Las Vegas)*, pages 263–274, April 1994.

8. L. Eikvil, T. Taxt, and K. Moen. A fast adaptive method for binarization of document images. In *First Int. Conf. on Document Analysis and Recognition (Saint Malo)*, pages 435–443, 1991.

9. C.S. Fahn, J-F. Wang, and J-Y. Lee. A topology-based component extractor for understanding electronic circuit diagrams. *Computer Vision, Graphics and Image Processing*, 44:119–138, 1988.

10. R.C. Gonzalez and R.E. Woods. *Digital Image Processing*. Addision Wesley, Reading, Massachusetts, 1992. ISBN 0–201–50803–6.

11. K.J. Goodson and P.H. Lewis. A knowledge-based line recognition system. *Pattern Recognition Letters*, 11(4):295–304, 1990.

12. M. Ilg. Knowledge-based understanding of road maps and other line images. In *10th IAPR Int. Conf. on Pattern Recognition (Atlantic City)*, volume 1, June 1990.

13. S.H. Joseph and T.P. Pridmore. Knowledge-directed interpretation of mechanical engineering drawings. *IEEE Trans. Pattern Analysis and Machine Intelligence*, 14(9):928–940, September 1992.

14. T. Kaneko. Line structure extraction from line-drawing images. *Pattern Recognition*, 25(9):963–973, 1992.

15. R. Kasturi, S.T. Bow, W. El-Masri, J.R. Gattiker, and U.B. Mokate. A system for interpretation of line drawings. *IEEE Trans. Pattern Analysis and Machine Intelligence*, 12(10):978–992, 1990.

16. H.P. Kramer and J.B. Bruckner. Iterations of a non-linear transformation for enhancement of digital images. *Pattern Recognition*, 7:53–58, 1975.

17. M.D. Levine and A. Nazif. Dynamic measurement of computer generated image segmentation. *IEEE Trans. Pattern Analysis and Machine Intelligence*, 7(2), 1985.

18. T. Nagao, T. Agui, and M. Nakajima. An automatic road vector extraction method from maps. In *9th IAPR Int. Conf. on Pattern Recognition (Rome)*, pages 585–587, Nov 1988.

19. V. Nagasamy and N.A. Langrana. Engineering drawing processing and vectorization system. *Computer Vision, Graphics and Image Processing*, 49:379–397, 1990.

20. N. Otsu. A threshold selection method from gray-level histograms. *IEEE Trans. Systems, Man and Cybernetics*, SMC-9(1):62–66, January 1979.

21. B. Pasternak and B. Neumann. ADIK: an adaptable drawing interpretation kernel. In *Avignon 1993, 13th Int. Conf. on AI, Expert Systems and Natural Language (Avignon)*, volume 1, pages 531–540, May 1993.

22. P. Puliti and G. Tascini. Interpreting technical drawings. *The Computer Journal*, 33(5), 1990.

23. T.W. Ridler and S. Calvard. Picture thresholding using an iterative selection method. *IEEE Trans. Systems, Man and Cybernetics*, 8(8):630–632, August 1978.

24. J. Serra. *Image analysis and mathematical morphology*. Academic Press, London, 1982. ISBN 0-12-637240-3.

25. M. Sonka, V. Hlavac, and R. Boyle. *Image processing, analysis and machine vision*. Chapman Hall, Cambridge, Great Britain, 1st edition, 1993. ISBN 0-412-45570-6.

26. S. Suzuki and T. Yamada. MARIS: map recognition input system. *Pattern Recognition*, 23(8):919–933, 1990.

27. O.D. Trier and T. Taxt. Evaluation of binarization methods for utility map images. In *First IEEE Int. Conf. on Image Processing (Austin)*, November 1994.

28. Y.J. Zhang and J.J. Gerbrands. Objective and quantative segmentation evaluation and comparison. *Signal Processing*, 39:43–54, September 1994.

Automatic Region Labeling of the Layered Map

Min-Ki Kim, Mun-Kyu Park, Oh-Sung Kwon and Young-Bin Kwon

Computer Vision Laboratory, Dept. of Computer Engineering
Chung-Ang University
221 Heuksuk-Dong, Dongjak-Ku, Seoul 156-756, Korea
e-mail: ybkwon@ripe.chungang.ac.kr

Abstract. In this paper, we describe an automatic region labeling method, which identifies each region and recognizes region names. Before tracing the region boundaries, it extracts the region names which consist of characters, dots, dashes, and indication lines. It uses two recognition methods to recognize characters in the region name. In the case of recognizing the isolated characters, it uses the open and close features. The characters touching boundaries are recognized by template matching. After removing the components of region names from a map image, the boundaries of each region are extracted. After which it then vectorizes the region boundaries. From these recognition results, the original map can be constructed. It reduces the storage to one fifth of the original data. The proposed method shows 95% accuracy of region labeling.

1 Introduction

GIS (Geographic Information System) generally consists of four parts: input, storage, analysis, and output. This paper covers the input part. Until recently, much of the data input was conducted manually, making time and cost a large requirement. Research on automatic map input has been conducted actively [1-4]. The main advantage of a computerized automatic input system is fast processing time. Suzuki accomplished a map input in 25 hours using MARIS (MAp Recognition Input System), which usually took 104 hours manually [3].

The purpose of this research is to improve the inputing time and storage by automatic recognition and vectorization [5,6]. The recognition results are stored in the form of data structure which can be easily processed. Scanned map images have noise and errors caused by scanned mechanisms, so we propose a system which corrects noise and errors. We use two methods to recognize the characters. The open and close features are used to recognize the isolated characters and template matching is used to recognize the touching characters on the region boundary.

2 Configuration of the Proposed System

2.1 Maps used

Three types of maps are used in this system. As seen in Fig. 1, the region name of Class-A only has single alphabet. In Class-B and Class-C, region names consist of alphabets, digits, and special symbols. Table 1 shows the possible character sets.

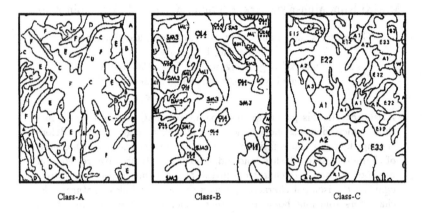

| Class-A | Class-B | Class-C |

Fig. 1. Parts of the layered map.

2.2 System overview

Fig. 2 shows the overview of the proposed automatic region labeling system. After loading the scanned image, it extracts characters and indication lines, then the extracted components are merged according to the geometric relation. When the region name and indication line are paired, the region name is moved into the region where the indication line points. Following this step, all the characters and symbols are extracted and saved. Now, only region boundaries remain on the map.

Table 1. Character sets of region names.

CLASS-A	(A|B|C|D|E|F|G|W|X)
CLASS-B	(W|X) or (GW|GP|GM|GC|SV|SP|SM|SC|ML|CL|OL |MH|CH|OH|PT|RK|EV|PS) + (0|1|2|3|4|5|6) + (underline|dash|dot)
CLASS-C	(H|J|K|L|M|N|W|X) or (B|G) + (1|2) or (A|I) + (1|2|3|4) or (C|D|E|F) + (1|2|3|4) + (1|2|3|4|5|6|7|8|9) + (underline|dash|dot)

Before it finds region boundaries, a thinning process is performed for simplifying the vectorization. The thinning process requires a lot of processing time. Each region boundary is extracted by contour tracing. One region boundary can include another region boundary. Before the region name is decided, the geometric relation of each region must be examined; followed by the labeling of the region. After labeling, the vectorized boundaries are stored on disk.

We register the character sets to be recognized. In the case of isolated characters, we use the size independent features. Therefore, we register only one model per character. However, touching characters are recognized by template matching, and we register multiple models for the same character.

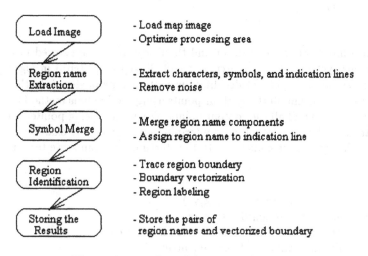

Fig. 2. An overview of the proposed system.

3 Implementation of the Proposed System

3.1 Contour tracing

To extract characters, symbols, and regions from the map image, we use the contour tracing method. In order to discriminate the character and inside area of the region boundary, we have to distinguish the interior contour and the exterior contour.

The following two algorithms describe the contour tracing method. Algorithm 1 describes how to find the starting point of contour tracing. Algorithm 2 shows how to trace the connected component.

Algorithm 1. Finding the starting point for contour tracing
- Initial value of image: Background0, Foreground3 (2-bit image)

```
- WIDTH, DEPTH: Image width and depth
- IMG(x,y): the value of coordinate (x,y) of the image
- Set image boundary to 0
1        FOR( y=1; y<DEPTH-1; y=y+1 )
2        BEGIN
3          FOR( x=1; x<WIDTH-1; x=x+1 )
4          BEGIN
5            IF( IMG(x,y)=0 AND (IMG(x+1,y)=2 OR IMG(x,y)=3))
6               OR (IMG(x,y)=1 AND IMG(x+1,y)=3) )
7               CALL Tracer( x, y )
8          END
9        END
```

Algorithm 2 traces the contour and the traced points are stored in a linked list. In this system, we use a circular double linked list. Step 3 in algorithm 2 saves the smallest y value for deciding the contour type in step 6. In the case of Fig. 3, we cannot find all the contour points using the Pavlidis method [7], so we propose step 5 to find all the contour points. When the tracing pointer arrives at the starting point, we do not simply terminate the trace but examine the next point. If the two points coincide with the starting two points, we terminate the trace.

Algorithm 2. Contour Tracer(x,y)
- x,y: Starting point for contour tracing
- clockwise tracing according to 8 directional chain code
- minp: minimum y value on boundary pixel

step 1 Check 8 direction for NP(next pixel). If it is found, set 0 to 1.

step 2 If NP is found, then Goto step 3, else Goto step 7.

step 3 If y value of NP < y value of minp, then modify minp.

step 4 If the value of NP is '3', then set to the value of NP to '2'. Store the chain code direction and the coordinate of NP.

step 5 Check NP is equal to starting point, and then examine the next contour pixel. If the two points coincide with the starting two points, then Goto step 6, else Goto step 1.

step 6 Decide the contour type according to Fig. 4.

step 7 Store the contour points and return.

Because algorithm 2 traces the next point clockwise, the exterior contours are traced clockwise at the highest point of contour boundary, and the interior contours are traced counterclockwise. This characteristic is used to decide the contour type (Fig. 4).

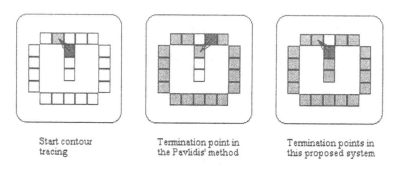

Start contour   Termination point in   Termination points in
tracing   the Pavlidis' method   this proposed system

Fig. 3. Conditions of contour tracing termination.

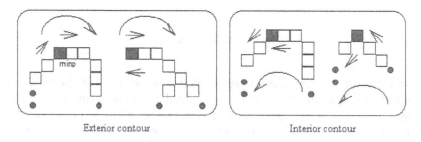

Exterior contour   Interior contour

Fig. 4. Types of contour decision.

3.2 Connection and disconnection of lines

The lines in a map image are easily disrupted by scanning errors, noise, and touching elements. Either a single line is disconnected from many lines, or many lines are connected together. To extract correct information from the map image, we have to solve these problems. To solve these problems we find the end points of the line using the interior angle which is computed from three evenly spaced points along the contour.

Using the information of the line thickness and the distance between two endpoints; disconnected lines are estimated. Then the connection is done by the Bresenham line drawing algorithm. Using line length, an indication line is estimated, and the touching indication line is disconnected.

3.3 Recognition of characters and symbols

1) Isolated Character Recognition

It is relatively easy to recognize the character which is not touching other elements. To recognize isolated characters, we use the Impedovo features [8] which are based on the background types. The background types are decided by a combination of open and close features in four outward directions. Fig. 5-(a) shows the background types of 'A' and Fig. 5-(b) shows various background types.

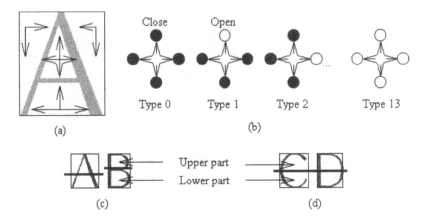

Fig. 5. Feature extraction.

To represent the Impedovo features effectively, we divided the minimum bounding rectangle into two parts; upper and lower part. Using a histogram we get a peak. If we have a peak in the middle part as in Fig. 5-(c), we partition the region with this peak. Otherwise we partition the region with the center line as in Fig. 5-(d). The proposed features, UF_i and LF_i, are defined as:

$$UF_i = \frac{number\ of\ white\ pixels\ in\ type\ i}{total\ number\ of\ white\ pixels\ in\ the\ upper\ part},$$

$$LF_i = \frac{number\ of\ white\ pixels\ in\ type\ i}{total\ number\ of\ white\ pixels\ in\ the\ lower\ part}$$

where, i=0~13. Comparing these features to each model, the recognition is completed to the class of minimum difference.

2) Touching Character Recognition

When characters are touching the boundary line, it is very difficult to find the characters without any prior knowledge. Therefore, we first recognize the isolated characters. Secondly, using the information acquired from the recognized characters, we estimate the location and size of the touching character.

In the case of Fig. 6-(a), at least one isolated charter exists. Thus we can set the search area and perform template matching in the area. The part shown in dotted line represents the search area. Moving the template of each model to the right and down, we measure the similarity to each model [Fig. 7]. If the similarity to a model is larger than the threshold, we stop and classify the character to the model. When all the characters are touching the boundary line as in Fig. 6-(b), we fail to recognize the characters because we cannot use any prior knowledge. This method needs a lot of processing time. However, using the lexical knowledge of region names, we can reduce the processing time effectively.

3) Symbol Recognition

Indication lines, dots, and dashes can be recognized by structural features.

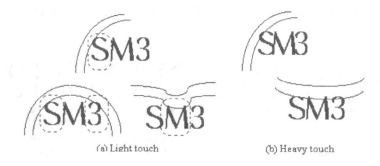

(a) Light touch   (b) Heavy touch

Fig. 6. Several examples of touching characters.

Fig. 7. Template matching within the search area.

Fig. 8-(a) shows the indication line which has a single connected component having two end points, and the directions of the two end points are opposite. However, the horizontal indication line can be considered as underline. Fig. 8-(b), 8-(c) is a dot and a dash, these two are discriminated by the height to width ratio.

4) Recognized Components Grouping

Recognized characters, dots and dashes are merged into a region name. We know the minimum bounding rectangle of each recognized component [Fig. 9-(b)]. We measure the horizontal neighbor with a threshold which is set by the characteristics of each component. Firstly, we merge the horizontal neighbor [Fig. 9-(c)]. Secondly, the merged dots and dashes are combined with upper character groups [Fig. 9-(d)]. Unmerged underlines remain as indication lines, and acquired

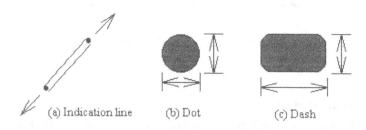

(a) Indication line   (b) Dot   (c) Dash

Fig. 8. Components of a region name.

region names are verified by comparing the possible lexical combinations in Table 1.

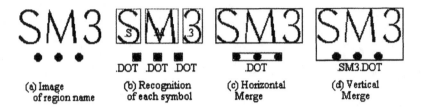

(a) Image of region name   (b) Recognition of each symbol   (c) Horizontal Merge   (d) Vertical Merge

Fig. 9. Merging process of the recognized components.

3.4 Vectorization of the region boundary

After extracting the components of the region name and indication lines, only boundaries of each region remain on the map image. To reduce error originating from boundary thickness we obtain the skeleton of region boundaries using the classic thinning method [7]. From this skeleton image, we trace the region boundaries using the method of section 3.1, and the traced boundaries are vectorized by iterative bisection method, which is exposed in Fig. 10. If the largest distance of line segment is greater than a threshold, we bisect the line segment recursively. Otherwise, we vectorize the segment.

When the boundaries of each region are vectorized, the vectorized results of the same boundary can differ [Fig. 11-(b)]. To eliminate this discrepancy we segment the boundary with the branching points. Then, we vectorize the boundary segments from the top-left point to the bottom-right point [Fig. 11-(c)]. The boundary without a branch is vectorized from the top-left point clockwise. After all, we can attain the same vectorized result with the two different contours which are traced in the opposite direction.

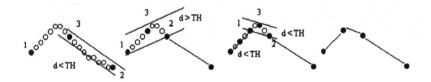

Fig. 10. Iterative bisection method.

3.5 Region labeling

A region is defined as a vectorized boundary, and its name is decided as a combination of recognition results of characters and symbols. Most of the region

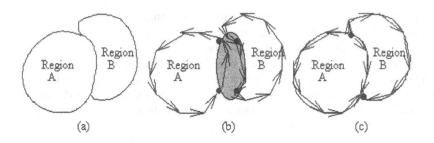

Fig. 11. Vectorization of the region boundary.

names are located inside the boundary. But when a region is small, the region name is outside the boundary and an indication line points to the region. In this case, we move the region name to the inside of the region. As illustrated in Fig. 12, when we extend the indication line toward the region, the mid point of the nearest two crossing points becomes the new location of the region name.

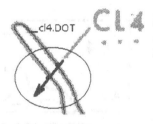

Fig. 12. Relocation of region names.

Therefore, a region is labeled by the region name it includes. When one region includes another region, the region labeling is done according to the geometric relation between the regions, that is, the labeling is done with the innermost region. After labeling, the region without a label is noted as unlabeled.

4 Experimental Results

Used maps are 400 dpi binary images of a size which is 24×30 inches (9,600 × 12,000 pixels). We need approximately 14 MB storage to load an image. The proposed system has been implemented in SUN SPARC station IPX with 130 MB swap partition.

Fig. 13 shows some results of the experiment. In (a), '1' in 'ML1' is touching a region boundary, but this system can extract '1' using the information of 'ML'. The area marked as a circle in (b) shows the connected results of a disconnected boundary. In (c), the indication line is touching the region boundary. It is correctly recognized using the information of end point and line length. Figure (d) and (e) show a correctly recognized general and complex indication line respectively. In (f), we fail to recognize the complex indication line because the two indication lines are not on a virtual straight line.

This system shows 95% accuracy of region labeling. Processing time is about 10 hours per map; 8 hours for thinning, 2 hours for the rest. The space for storing the region labels and vectorized boundaries is about 3 MB. Most of the incorrect recognitions come from touching characters and indication lines.

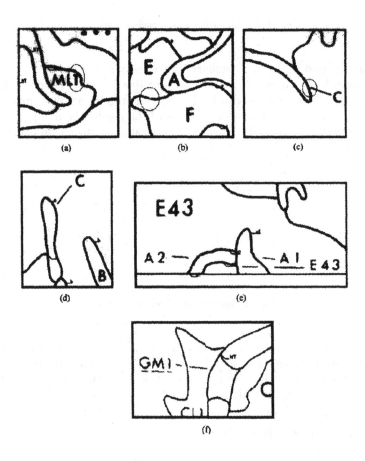

Fig. 13. Results of recognition.

5 Conclusions

We have developed an automatic region labeling method. In our research, we have proposed a contour tracing algorithm and a vectorization method. Most of the characters are not touching a boundary line, however some of the characters are touching a boundary line. To deal with these two different states, we have applied two different character recognition methods. We have used the open and close features to recognize isolated characters and used a template matching method to recognize touching characters.

The proposed method can process the map in 10 hours with 95% accuracy. In addition, it has reduced the storage to one fifth of the original data. When the job was done manually, it took one man more than a week. In further study, we need to develop a more effective recognition method of touching characters to improve recognition accuracy. To reduce the processing time we also need to do a study of fast and accurate vectorization method without thinning.

Acknowledgements

This work was supported by KOSEF (A Study on the Image Analysis Technique for Extension of the ICARE System).

References

1. Y.B. Kwon, J.R. Lee, Implementation of an automatic recognition system for extracting the map information, proc. of the 6-th IPU workshop, pp. 116-122, 1994.
2. R. Kasturi et al, Map Data Processing in Geographic Information System, IEEE Computer, Vol. 22, No. 12, pp. 10-21, 1989.
3. S. Suzuki and T. Yamada, MARIS: Map Recognition Input System, Pattern Recognition, Vol. 23, No. 8, pp. 919-933, 1990.
4. M. Ejiri et al, Automatic Recognition of Engineering Drawings and Maps, in Image Analysis Applications (R. Kasturi ed.), pp. 73-126, Marcel Dekker, 1990.
5. J.M. Ogier et al, Attributes Extraction for French Map Interpretation, Proc. of the ICDAR93, pp. 672-675, 1993.
6. Rik D.T. Janssen, Robert P.W. Duin, Albert M. Vossepoel, Evaluation Method for an Automatic Map Interpretation System for Cadastral Maps, Proc. of the ICDAR93, pp. 125-128, 1993.
7. T. Pavlidis, Algorithm for Graphics and Image Processing, Computer Science Press, pp. 142-148, 195-203, 1982.
8. S. Impedovo, A New Method for Automatic Reading of Typed/Handwritten Numerals, Proc. of IWFHR II, pp. 427-433, 1992.

Verification-Based Approach for Automated Text and Feature Extraction from Raster-Scanned Maps

Gregory K. Myers, Program Director
Prasanna G. Mulgaonkar, Director
Chien-Huei Chen, Senior Computer Scientist
Jeff L. DeCurtins, Computer Scientist
Edward Chen, Research Engineer

SRI International
Information, Telecommunications, and Automation Division
333 Ravenswood Avenue
Menlo Park, California 94025
Phone: (415) 859-4091; Fax: (415) 859-5510
E-mail: myers@erg.sri.com

Existing systems for converting maps to an object-oriented form suitable for a geographic information system (GIS) are only partially automated. Most published approaches for automated interpretation of raster-scanned maps assume that the map is composed of various graphic entities, and that the vast majority of pixel positions on the map each belong to only one type of graphic entity and can therefore be geometrically segmented. However, complex color topographic maps contain several layers of information that overlap substantially (often within a single color plane), making it impossible to geometrically segment the map data into distinct regions containing a single class of graphic object. Here we describe a verification-based approach that uses various knowledge bases to detect, extract, and attribute map features without requiring the presegmentation of graphical entities. This approach builds on SRI International's (SRI's) verification-based computer vision and character recognition methodologies. The approach can also be applied to other types of documents containing a mix of text and graphics, such as engineering drawings, electrical schematics, and technical illustrations.

Keywords: Automated Map Interpretation, Text Extraction, Map Feature Extraction

1 Background

Many organizations have large sets of hard-copy graphic documents, including maps, engineering drawings, electrical schematics, and technical illustrations, that are used routinely in their operations. Recent advances in computer technology have permitted graphic information to be stored and accessed more conveniently and cost-effectively in electronic form than on paper. If such information is represented as distinct graphic entities that correspond to the logical components of real objects, the information can be presented and manipulated in a highly productive manner. In the case of topographic maps, therefore, we desire the ability to convert the graphic information to a form suitable for a geographic information system (GIS) database with attributed features and associated text labels.

Existing commercial systems for map data conversion are only partially automated. In general, feature-separate or color-separate data are extracted from scanned hard-copy maps and processed individually to extract areal features (e.g., bodies of water); point features (e.g., buildings, bridges); and linear features (e.g., grids, roads, railroads, streams, contours). Automated recognition of areal features and point features in feature separates or color separates generally works reasonably well, because the features are usually represented by symbols or homogeneous areas of screen patterns that are physically isolated from other features on the map or in other separate layers.

The extraction of linear features and text, however, is currently performed manually for several reasons. Various types of linear features often appear within a single color plane, and they touch, cross, and overlap. Although segments of linear features (continuous, with no branches or crossings) can be detected and vectorized by automated processes, manual methods are currently used to identify and attribute a particular type of linear feature, and to link individual segments of a linear feature into a continuous entity (e.g., a pipeline that goes underneath a roadway).

Commercially available character recognition methods are not designed to cope with conditions of text that appear on maps. Most character recognition methods require manual definition of the zones of text among the nontext information, and rely upon having straight, horizontal lines of text within each zone. Many map labels consist of only a few characters; hence, it is difficult to distinguish them from other map items, and it is more difficult to establish their orientation and group them into words and labels. The text labels for some natural features (rivers, mountain ranges) and some linear features (railroads, canals) can be oriented at any angle, and the alignment of characters in the label can curve. Characters can touch and cross other map features. Place names may be hyphenated, and can be in a combination of type styles (roman and italic).

The entire process of associating labels with the features they reference is currently performed manually. Automating this process is a major challenge, because of the high density of information on the map, and because generalization (displacing features and labels, smoothing features, and aggregating features based on scale and symbolization) is highly dependent on the specific feature-to-feature relationships, scale, and use of the map.

This paper describes an approach called *verification-based recognition* for converting map documents from raster-scanned images to object-oriented form. This approach uses contextual knowledge and constraints in a novel way to interpret graphic and text information in the document, adapting SRI's existing verification-based computer vision and character recognition methodologies.

2 Verification-Based Approach

2.1 Verification-Based Recognition

Published approaches for automated interpretation of raster-scanned maps [1; 2; 3; 4] have generally assumed that maps are composed of various graphic entities, and that the vast majority of pixel positions on the map each belong to only one type of graphic entity and can therefore be geometrically segmented. As a result, the processing steps in these approaches have generally been to (1) isolate each of the graphic entity types (e.g., lines, text), (2) apply entity-specific recognition, and

(3) merge the results into an integrated interpretation by applying some contextual knowledge. However, complex color topographic maps contain several layers of information (e.g., hydro, terrain, transport, elevation, political, urban areas) that overlap substantially (often within a single color plane), making it impossible to geometrically segment the map data into distinct regions containing a single class of graphic object. Because the graphic characteristics of information in each of these layers differ, an approach that applies the same processing uniformly to features in all layers may not yield optimum results. For example, the parameter values that are appropriate for controlling a vectorization and linking process for one type of linear feature (e.g., roads) may not work well on another type of linear feature (e.g., streams).

SRI has started a preliminary investigation of verification-based recognition approaches that are expected to be much more successful, because they take advantage of the general principle that objects and their relationships can be recognized much more successfully if algorithms know what to look for and do so in a directed search. A verification-based approach uses contextual knowledge and constraints to formulate and then verify interpretation hypotheses. In contrast, conventional image-recognition approaches try to identify objects from pixels first, and only afterward apply any contextual knowledge. By interpreting the pixels in terms of the possible hypotheses, verification-based approaches should have several important advantages:

- They will operate successfully amid extraneous graphic information, even where the graphical object of interest is touching or overlapping other information. Recognition techniques that depend on having the object of interest isolated, or the unwanted background "removed" from the image, may not be as successful, because an error in the isolation or removal process will preclude success in the subsequent processing steps.

- They are better suited to taking advantage of the rich a-priori knowledge associated with most maps. Information from legends, map specifications, gazetteers, and existing digital databases can be a guide to the map data and can drive the interpretation process.

- They are better suited to take advantage of the interrelationships between information expressed in different graphical modes. For example, the presence of a bridge symbol can bolster the hypothesis that two linear water segments located on either side of a road should be linked.

- A verification-based approach is amenable to extracting only the data that are of interest. Therefore, less processing time will be required than approaches that try to interpret everything on the map. This capability is especially valuable when different information is desired from several types of maps covering the same geographic area. In addition, this capability can be part of an interactive data extraction tool that allows a user to search, or "query," the raster map data for specific information.

In this approach, models of map features from scanned images of legends and map specifications would be constructed in an off-line training process. Linear, point, and areal features can be detected and extracted by applying object recognition techniques to generate and then verify hypotheses about the features. We can use the recognized features and lexicons of place names obtained from a gazetteer to hypothesize the position and content of text labels that appear on the map, and then attempt to recognize the text. Automated association of the text label with the map feature will be an implicit part of this process.

Instead of treating the extraction of features and text, tagging, and association as independent and noninteracting processing stages, our approach integrates these processes more tightly to handle the uncertainties and resolve the ambiguities of each processing stage. This can be accomplished by having the initial recognition of features produce hypotheses instead of making hard recognition decisions. Then, feature hypotheses can *drive* the search for particular text labels; similarly, the hypothesis of a text label can *drive* the additional search for the feature with which it should be associated. In other words, both the detection and the association of logically related map components are confirmed simultaneously when a consistent interpretation of *all* of the data is found. For example, the presence of a school is indicated by a school symbol and an associated text label. Suppose the extraction of text yields a hypothesis containing the word "School," but no corresponding school symbol was found near that word. The detector for the school symbol could be rerun in the area near the word with parameters set to make the detector more sensitive. This approach therefore involves a combination of bottom-up and top-down processing, which has been found to be a useful control strategy in other map interpretation work involving low-level recognition processes and high-level knowledge bases [3; 5].

2.2 Automated Map Data Extraction Architecture

An overview of an architecture for automated map data extraction that incorporates a verification-based approach is shown in Figure 1. It operates on scanned raster map data. The output consists of a data structure containing features, attributes, and confidences in a format suitable for interfacing to a GIS for evaluation and editing.

The architecture consists of a set of graphic data extraction modules and an extraction controller. The data extraction modules extract point, linear, and areal features and text from the raster map data. Each data extraction module is designed to look for specific map data amid the other information that may be present in the raster image. Legends and map specifications are used to generate models of the features and text styles that are expected to be on the map. The process does not depend on first segmenting different types of data (such as text and nontext data) into separate images. The resulting features and their associated attributes and confidences are initially treated as hypotheses.

The extraction controller exploits the interrelationships between features and the relationships between the features and already-existing knowledge bases to detect errors or incomplete data and initiate further processing to resolve them. The extraction controller refines and augments the hypotheses generated by the data extraction modules in an initial pass. A conflict resolution module detects errors by examining the consistency of each new feature hypothesis that is generated by the data extraction modules with associated information sources. In addition to detecting the incompatibility between features within a single map, it compares the position of features with auxiliary GIS data, such as lower-resolution map data already entered and verified, or existing electronic databases. Hypothesis generation is invoked in cases where associated information is missing or additional processing may be able to resolve a conflict or confirm a low-confidence feature hypothesis. For example,

- Text and feature hypotheses is generated from gazetteer data and from features found in auxiliary GIS data covering the same area.
- Unrecognized text is matched with features for which the corresponding text has not been found.

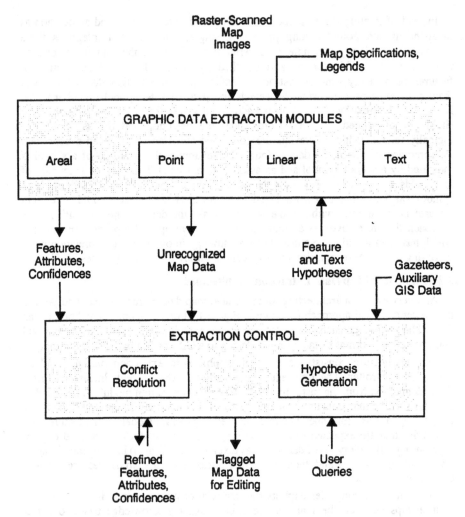

Figure 1. Automated Map Data Extraction Architecture

- Linkages are proposed for the end points of linear features based on the presence of other nearby feature hypotheses.

Our initial work, described in the following sections, consists of demonstrating the verification-based approach for various data extraction modules by adapting and applying existing verification-based techniques developed for other image domains. Full development of all data extraction modules and extraction control modules has yet to be completed.

3 Verification-Based Data Extraction

Below, we present three examples of the power of a verification-based approach to the interpretation of raster-scanned map data. Each of the examples contains a technique that is focused on extracting a specific type of information from the map, and each of the techniques works amid the clutter of the remaining graphic information on the map. These techniques were originally developed for other computer vision and document understanding applications in which a noisy and cluttered image environment required verification-based approaches.

3.1 Map Symbol Feature Extraction

Point features on a map are represented by two-dimensional symbols, and areal features are represented by regions defined by either a fine print screen pattern or a repeated pattern of a discrete symbol. In many cases, these symbols are touched or overlapped by other map entities. Figure 2 shows part of a text image, from a USGS topographic map, that contains several marsh symbols. Notice that there is appreciable variation in the shapes of the printed symbols.

To this test image we applied object recognition techniques developed by SRI for industrial computer vision domains [6; 7; 8]. In these techniques, training is performed automatically by showing it objects in sample images. Geometric shape features of the objects are extracted and ranked automatically according to how reliable they are relative to image noise and clutter, and how well they contribute to uniquely distinguishing one object from another. During recognition, key features are detected in the image and trigger the generation of hypotheses about the presence and location of particular objects. The hypotheses are verified by checking for the presence of the remaining features in the object model. This method can recognize objects in spite of partial occlusion or noise in the image data. This approach is similar to the hypothesis-and-test approach taken in [9; 10], except that because our training procedure uses multiple samples of symbols extracted from actual map images, the

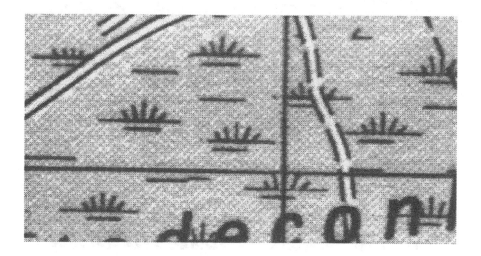

Figure 2. USGS Topographic Map

reliability of the feature detection process and the effects of image degradations can be factored into the matching process: features that have been more often successfully detected in the training set receive a higher weight during the matching process.

The features in the model of the marsh symbol consisted of the relative positions, angles, and lengths of its seven line segments. In a test image that contains 56 marsh symbols, this verification-based recognition approach recognized 42 of them, with one false recognition. In Figure 3, marsh symbols that have been recognized are marked off by rectangles. Notice that many symbols are correctly recognized, in spite of the variation in their shapes and occlusion by other map entities, and that most of the unrecognized symbols were only partially visible.

Figure 3. Test Image with Recognized Symbols

3.2 Linear Feature Extraction

Our approach for linear feature extraction is based on adapting SRI's technique for line delineation in imagery [11; 12]. The first step in this approach consists of identifying pixels in the image that appear to be a part of some linear segment. These pixels could be part of a thin linear feature such as a stream or railroad, or a centerline of a thicker linear feature, such a divided highway. This operation is performed with a specially designed set of operators that adaptively adjust their parameters to the image content.

The second step is to link the detected line pixels to form a list of linear segments. This step applies generic criteria about linear features (continuity, contiguity, coherence, length) to locate perceptually obvious lines, independent of the type of linear feature being processed. The processing in this second step primarily consists of a clustering or association step based on geometric proximity, followed by the construction of a Minimum Spanning Tree (MST) through the points of each cluster.

The third step applies semantic criteria specific to the type of feature to be detected. Such criteria would include color, thickness, curvature per unit distance, branching behavior, and distinguishing marks, such as periodic tick marks on railroads and the spacing and length of dashed lines. The semantic linker modifies the link structure and filters out linear segments that do not qualify. The link structure is converted to vector representation by fitting lines or parameterized curves to the linked list of points.

In comparison with other raster-to-vector conversion methods, this approach separates the use of geometric properties of inherent linearity from the use of semantic information that varies for each type of linear feature, whereas other methods apply the same processing and same criteria to all line-like objects in the image. Figure 4 is an example of linear delineation applied to road extraction in aerial imagery. Figure 5 shows the results of applying the same algorithm to a street map. Because the MST that links the neighboring line pixels does not require strict connectivity, gaps in the lines representing streets can be correctly linked. In addition, in most cases the street lines are successfully found in spite of characters touching the street line.

3.3 Text Extraction

An example of a verification-based approach for text labels is shown below in Figures 6–10. We used a gazetteer to automatically generate hypotheses of the characters in the labels and their approximate locations in a 1:24,000-scale, raster-scanned USGS topographic map. The figure shows a small section of the black image plane extracted from the map. Some text labels are touching graphic data, and because the process that extracts the black pixels from the scanned map image is imperfect, some text is partially obscured by noise. A lexicon of text names likely to appear on the map was obtained from a gazetteer. Zones of text were detected in the image and tested against hypothesized labels. Our approach for finding text zones searches for patterns of short run lengths that are characteristic of text. The areas with detected text-like characteristics are then linked into contiguous zones by various criteria for co-linearity. This process requires two passes through the data—one for detecting horizontally oriented text, and the other for detecting vertically oriented text. In contrast to [13; 14], our approach for finding text zones is not based on finding individual characters, making it easier to link the zone across intermingling non-text data. Figure 7 shows the result of this process for horizontally oriented text. The rectangles indicate the presence of text zones. In this example the algorithm found 56

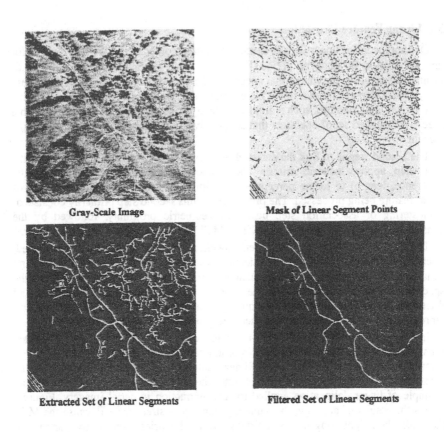

Gray-Scale Image   Mask of Linear Segment Points

Extracted Set of Linear Segments   Filtered Set of Linear Segments

Figure 4. Example of Road Delineation

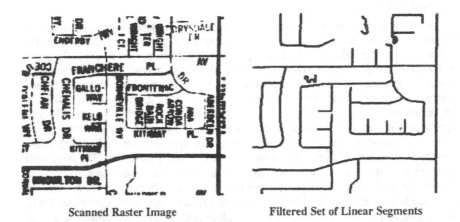

Scanned Raster Image   Filtered Set of Linear Segments

Figure 5. Example of a Road Delineation on a Street Map

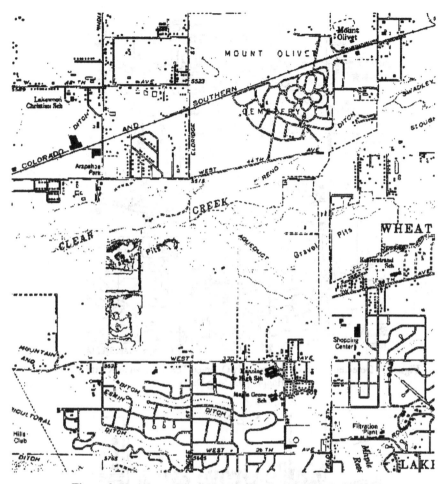

Figure 6. Black Image Plane Extracted from a 1:24,000-Scale USGS
Topographic Map

out of 65 of the text zones correctly. Other areas (not shown) containing non-text are also found, but these are discarded in the text verification step below. Currently the algorithm is not scale-independent and therefore did not merge some strings of text consisting of larger font sizes; this could be resolved in the future by adaptively setting the linking parameters according to the size of the detected areas and the map specifications.

The text zones define the areas where the text verification process operates. Each zone is first deskewed, and then linear-like features are removed from the zone. Figure 8 shows the effects of the linear-like feature removal process on one of the text zones. We then apply SRI's approach for keyword recognition [15], which detects the presence of a limited vocabulary of words based on whole word shape. The set of hypothesized text strings enables us to take advantage of the context for the whole label instead of just independently recognizing isolated characters, and the likelihood of correct recognition of the complete text string on the map can be increased

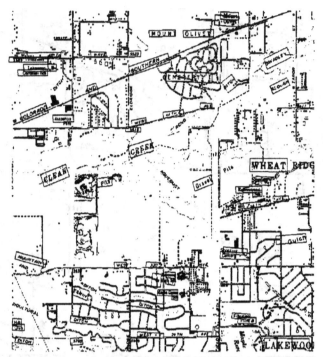

Figure 7. Detected Zones of Horizontally Oriented Text (Shown as Rectangles Overlapping the Map Data)

Text zone with linear-like features

Text zone after removal of linear-like features

Figure 8. Example of Removal of Linear-Like Features from Text Zone

enormously. Features in each text zone, such as ascending and descending strokes, curves, and spaces, are extracted and compared with those extracted from the hypothesized text labels. Figure 9 shows the results of text verification. Of the 56 text zones successfully located in this map section, 29 were correctly recognized using this method. Figure 10 shows close-up views of three text labels that were successfully detected and recognized, in spite of the labels' coarse resolution and the interference in the black image plane that is intermingled with the text. Because this method uses only word-level features, short text labels are more difficult to recognize. By augmenting this method with the ability to extract features of individual characters and recognize them, performance would improve on some of these short text labels.

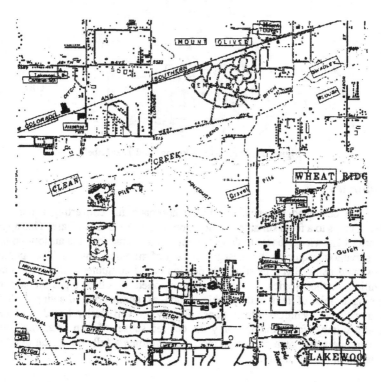

Figure 9. Zones of Verified Text (Shown as Rectangles Overlaying the Map Data)

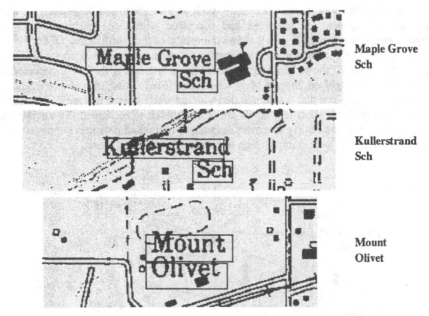

Maple Grove
Sch

Kullerstrand
Sch

Mount
Olivet

Figure 10. Successfully Recognized Text Labels

4 Summary

We have described a verification-based approach for converting raster-scanned images of maps to a form suitable for incorporation into a GIS database. This approach uses various knowledge bases to drive the detection, extraction, and attribution of map features, without requiring the presegmentation of graphical entities. The approach is demonstrated using SRI's existing verification-based computer vision and character recognition methodologies. It can also be applied to other types of documents containing a mix of text and graphics, such as engineering drawings, electrical schematics, and technical illustrations. Finally, a verification-based approach is especially suitable for extracting only the desired subset of map data, and could be used in an interactive data extraction tool, thus avoiding a full conversion of all of the information in the document.

5 References

1. Boatto, L., V. Consorti, M. Del Buono, S. Di Zenzo, V. Eramo, A. Esposito, F. Melcarne, M. Meucci, A. Morelli, M. Mosciatti, S. Scarci, and M. Tucci. 1992. "An Interpretation System for Land Register Maps," *IEEE Computer*, pp. 25–33 (July).

2. Consorti, V., L.P. Cordella, and M. Iaccarino. 1993. "Automatic Lettering of Cadastral Maps," *Proceedings of the Second International Conference on Document Analysis and Recognition*, Tsukuba Science City, Japan, pp. 129–132 (20–22 October).

3. Maderlechner, G., H. Mayer, and C. Heipke. 1993, "Conversion of Scanned Cartographic Maps to Geographic Information Systems using Semantic Models," *Proceedings of the Second Annual Symposium on Document Analysis and Information Retrieval*, pp. 339–347.

4. Suzuki, S. and T. Yamada. 1990. "MARIS: Map Recognition Input System," *Pattern Recognition*, Vol. 23, No. 8, p. 919.

5. den Hartog, J.E., T.K. ten Kate, and J.J. Gebrands. 1995. "Knowledge-Based Segmentation for Automatic Map Interpretation," *Proceedings of the International Workshop on Graphics Recognition*, University Park, Pennsylvania, pp. 71–80, (10–11 August).

6. Bolles, R.C., and R.A. Cain. 1982. "Recognizing and Locating Partially Visible Objects," *International Journal of Robotics Research* 1, pp. 57–82 (Fall).

7. Chen, C.H., and P.G. Mulgaonkar. 1992. "Automatic Vision Programming," *CVGIP: Image Understanding*, Vol. 55, No. 2 (March).

8. Gleason, G.J., and G.J. Agin. 1979. "A Modular Vision System for Sensor-Controlled Manipulation and Inspection," *Proceedings of the Ninth International Symposium on Industrial Robots*, Washington, D.C. (March).

9. Shimotsuji, S., O. Hori, and M. Asano. 1994. "Robust Drawing Recognition Based on Model-Guided Segmentation," *Proceedings of Document Analysis Systems*, pp. 353–376.

10. Shimotsuji, S., S. Tamura, and S. Tsunekawa. 1989. "Model-based Diagram Analysis by Generally Defined Primitives," *Proceedings of Scandinavian Conference on Image Analysis*, pp. 1034–1041.

11. Fischler, M.A., and H.C. Wolf. 1983. "Linear Delineation," *Proceedings of the IEEE CPR-83*, pp. 351–356 (June); also, *Readings in Computer Vision* (M.A. Fischler and O. Firschein, eds.), Morgan Kaufmann, pp. 204–209.

12. Fischler, M.A. 1994. "The Perception of Linear Structure: A Generic Linker," *Proceedings of the ARPA Image Understanding Workshop*, pp. 1565–1579.

13. Hontani, H., and S. Shimotsuji. 1995. "Character Detection Based on Multi-Scale Measurement," *Proceedings of ICDAR'95*, pp. 644–647.

14. Pierrot-Deseilligny, M., H. LeMen, and G. Stamon. 1995. "Characters String Recognition on Maps, a Method for High Level Reconstruction," *Proceedings of the Third International Conference on Document Analysis and Recognition*, Montreal, Canada, Vol. 1, pp. 249–252 (14–16 August).

15. DeCurtins, J.L. 1995. "Keyword Spotting via Word Shape Recognition," *Proceedings of the SPIE Symposium on Electronic Imaging*, San Jose, California, Vol. 2422 (February).

Software System Design for Paper Map Conversion

A.W.M. Smeulders[1] T. ten Kate[2]

[1] Intelligent Sensory Information Systems
Faculty of Mathematics and Computer Science
University of Amsterdam, Kruislaan 403
1098SJ Amsterdam, the Netherlands
smeulder@fwi.uva.nl
[2] TNO Institute of Applied Physics, P.O. Box 155, 2600 AD Delft,
the Netherlands

1 Introduction

It has been estimated that the conversion of graphic documents to electronic documents requires 15,000 years of work in the Netherlands alone.

We are engaged in a large scale effort to built a real system for the conversion of utility maps from paper to electronic GIS files. See figure 1 for a typical example of gas pipes. To this end, we have redesigned, restructured and implemented a graphical engineering drawing recognition system, starting from the ROCKI system [TenKate, DenHartog]. This system reached knowledge based interpretation of utility maps on the basis of decomposition of the engineering drawing in graphical primitives with subsequent contextual reasoning in a semantic network. The knowledge of the map was explicitly stored in this semantic network, also providing some repair mechanisms by local resegmentation of the image when a symbol recognizer fails. An OCR subsystem reads the values of printed characters and recognizes heads of arrows. The system was evaluated on realistic data. For a detailed account, see the references. In its aimed functioning it has as its competitors, for example [Joseph, Kasturi, Suzuki].

This communication presents considerations for the redesign of the system into a twin- system for flexible graphics document conversion. This communication barely discusses methods behind key components of the conversion process. It concentrates on structuring the software in well-defined components. Such considerations are not found in the above references as these papers concentrate on the rationale behind the algorithms how to perform a (part of the) graphics recognition task.

The first of the twin systems, nicknamed ROCKI VII to top the Hollywood movie series, performs the conversion of digitised paper utility maps to GIS-files. The process is aimed to run under supervision of the operator telling it where to analyse the scan and if necessary assist in its reasoning by providing interactive hints. The result is geometrically moulded on the digital base map of the Netherlands and stored away in a GIS, in our case the Small World system. It is described in more detail in section 2.

The twin system, nicknamed BESSI for no translatable reason, is a algorithm designer's tool used to validate graphic symbol recognizers on a database of examples. The examples are gathered with an interactive input tool from the digital scans and stored in a database. This system is described in more detail in section 3.

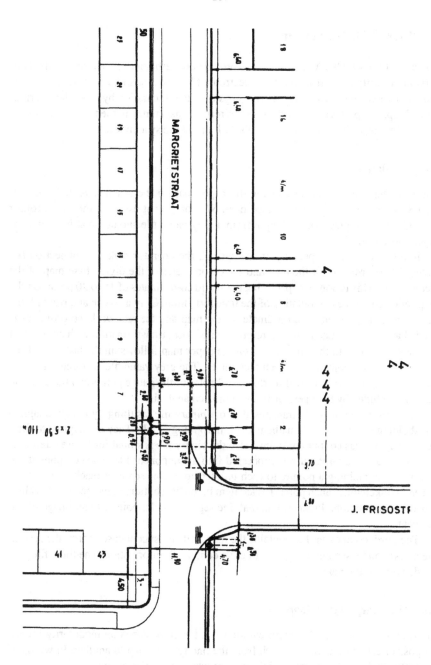

Fig. 1. Section of a typical utility map (displayed at reduced resolution). Data courtesy PNEM.

2 Rocki VII: the system

The aim of the ROCKI VII system is to provide cost efficient procedures for the conversion of utility maps from paper to electronic files. To that end, the system design is broken in three steps, see figure 2. The steps are: data separation by type, the interpretation loop, and preparation for the GIS. To reach transparency in the functioning of the system, in this design there are no feedback loops at the system level.

2.1 The data step

The data step starts with introducing the paper map to the scanner. A 400 dots per inch 8 bit grey value file is produced here. We consider it essential that for a robust recognition of graphic symbols input from a grey value file should be at hand in case binary processing fails.

In introducing the paper map in the system, the operator makes a connection between the co-ordinates of the map and the co-ordinates of the digital base map of the Netherlands. This is done by specifying the world co-ordinates of 15 to 20 points on the map followed by the computation of the transformation. Only a coarse and rigid global transformation between the co-ordinates of the map and the real world co-ordinates is needed at this point. Later on a more precise and local transformation with adjustment almost everywhere to the information on the paper map follows, in the last step of the analysis when all objects have been recognised. The grey value file is stored for future reference together with the real world co-ordinates of the gauge points and the areas on the map excluded by the operator from automated analysis.

The grey value image is transposed into a binary file resulting by locally adaptive thresholding the image with mathematical morphology tools. They are applied to filter out the larger areas containing stains. Also the binary file is stored for future reference.

The binary file on its turn is processed into a component file. The component file is a list file containing a pointer to each uninterrupted line segment (each segment between two skeleton branch points). Each item in the list contains some geometric values (position, orientation, thickness) on that line segment and a pointer to an image of the segment.

The third, component file contains the map in its most condensed form, the second file is less condensed and the grey value file contain all information. The three files are handed over to the next step.

2.2 The interpretation loop

In the system design of the interpretation loop, the emphasis is on modularity for the purpose of ease of making a switch from the one type of map to another. In words of the software designer, the reuse of the system was our prime concern.

To that end, we have enforced a strict separation of the graphical recognition software in detector modules, one for each graphic symbol, and reasoning modules. There is only very little software needed to connect these modules; each statement is localised for its purpose. In such a way, orthogonality of the code is ensured. If one wishes to replace a graphical symbol detector module by another one detecting a different style

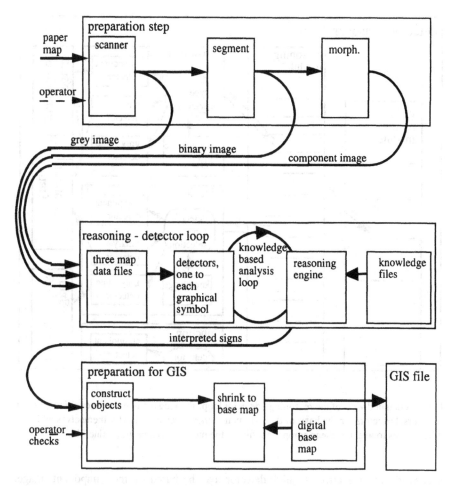

Fig. 2. Major components of the ROCK VII system. Note that the preparation loop serves to obtain several data representations of the map under analysis. No interpretation at this stage. Note also, in the reasoning loop there is no access to the data arrays than by means of one of the detector modules; one to each graphical sign. The data are hidden from the remainder of the analysis process. Also, detector modules are connected to the remainder of the analysis as part of a reasoning statement issued by the reasoning engine. In this way, the system lay-out is guaranteed to be of maximum modularity.

of drawing or another graphical symbol on a slightly type of map, one can do so while leaving the remainder of the software system unaffected, see Figure 3.

Access to the three different data representations is through detectors and detectors only. A detector is an algorithm for the detection of a specific graphic symbol, for example an arrow. It will report back presence or absence, sometimes also parameter values such as length and orientation augmented by certainty of observation. To all detectors the three data files are available on which a variety of detection techniques

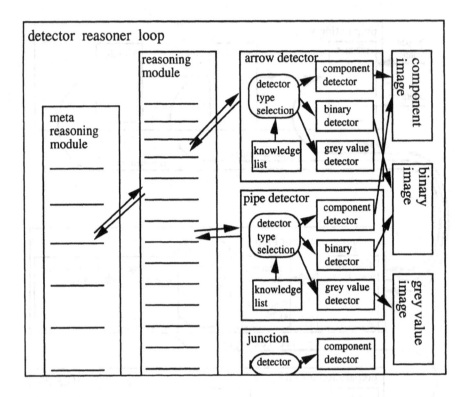

Fig. 3. A detailed view of the reasoning - detector loop, showing the interior functioning of a few detectors. Detectors, on the basis of the information they receive decide for themselves which of the data files to use: the cheap component file, or the more expensive grey value file.

may work. For the arrow a quick detector may be based on the component image, whereas a more accurate but costly procedure works on the grey value image. A detector module may consist of several of these detector techniques. Depending on the findings, the module will decide for itself on which technique to use. Usually that detector will be selected first which is computationally efficient. In descending order: component typed, binary typed, grey value typed. Detectors and connection patterns are designed from an axiomatic approach listing the desired properties of the algorithmic method on forehand, see the contribution to ICDAR [Jonk].

Detector modules are defined on a semantic level. A detector module has a (small) knowledge file to keep specifics for the drawing style at hand and a (small) internal reasoning module to decide between detection techniques. In this respect the detector module fulfils the paradigm of Object Orientedness in that data and knowledge in encapsulated and hidden from the remainder of the system. Such makes stand alone testing and performance evaluation possible.

The reasoning module is currently under reconstruction. It is based on a semantic network only allowing for strict reasoning steps. It will be replaced by Bayesian belief networks allowing for inexact reasoning as well.

2.3 Preparation for GIS

After a successful interpretation, in the third step preparation will commence for entering the GIS. With that information and on the basis of the list of interpretations resulting from step two, a GIS-file is formed.

At this point the connection between the object of the paper map and the real world co- ordinates of the digital base map of the Netherlands is refined. The 15 to 20 points introduced at the beginning of the analysis are a sparse transformation matrix field. At this point of the analysis, the matrix is filled in into a dense transformation matrix indicating for each point where it is positioned on the map.

When all is done, the conversion is verified by the operator and stored away as a GIS - file.

3 The BESSI tool

The BESSI tool is twin system of software tools for the tuning of map domain specific detectors. It is based on the image processing package SCIL-Image [Van Balen] and the commercially available database Illustra, formally known as Montage.

For each new domain of maps, a list of examples is made of all graphical symbols. Then, for each graphical symbol a detector is designed. Its parameter values are tuned, and its performance is measured with the aid of the BESSI tool, see fig. 4 for an overview.

BESSI awaits in the following input:

1. The software of the detector module to evaluate.
   Note the virtue of making a data- and knowledge encapsulated module in the ROCKI VII system, as the module can now be tested as one whole.
2. A dataset of test images containing examples the symbol's picture as well as counter examples thereof.
   To compose the dataset, BESSI contains an interactive image selection tool. Test images will be stored in the database contained in the BESSI tool.
3. Per test image the true identity of the graphical symbol.
4. An algorithm how to evaluate the quality of the image.

This can be a quadratic error, counting true / false, segmentation error by taking the EXOR between the ideal and the result, and alike.

When running the BESSI tool, evaluation results on the performance of the detector module are stored away for each image in the database of the test set together with the image data for inspection in retrospect. The BESSI tool is operated by a visual program [Koelma]. In its full extent, the BESSI tool will also include some image detoriation models, as in [Baird], so as to test the performance of the algorithm under a wide variety of circumstances.

The next generation of the BESSI tool will integrate a coupling of the SCIL-Image image processing package with the database system Illustra and a statistical library like S, a visualisation tool as GNU-Plot to serve as an integrated algorithm developers platform.

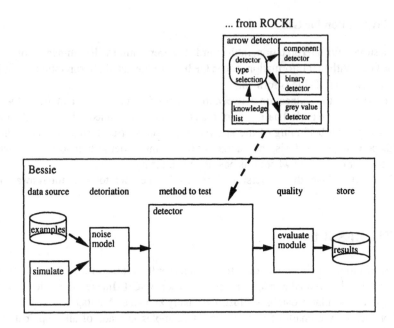

Fig. 4. BESSI system layout, testing an arrow detector.

4 Conclusion: methods engineering

In this paper we have discussed the redesign of a successful system, ROCKI, for paper map to digital file conversion. The redesign's prime target was to make the system modular, data encapsulated and reusable for the purpose of quickly learning of new map domains.

In the new system, ROCKI VII, transparency and maintainability are achieved by enforcing a strong separation between data driven actions and knowledge based reasoning. Quick learning is achieved in principle by strong separation of all symbol related actions into self contained detectors. There is one detector module for each graphical symbol, accessing one or more image data types to assess the actual presence or absence. Detectors render an assessment of their confidence of their answer. By encapsulating all data and all knowledge needed for the detection of one graphic symbol in one module, the way is opened to assess and optimise the performance of each detector by the BESSI system. In this way reliable, large scale systems can be build.

We view of these systems for methods engineering (as addition to software engineering) as a contribution towards making a designer science [Haralick, Baird] of what sometimes appears to be the accidental field of image processing.

5 Acknowledgement

Research sponsored in part by TopSpin - PNEM.

6 References

[Baird] Baird: Proc. ICDAR-93, ed. K.Yamamoto, IEEE-Press, Los Alamitos, 1993, 62-67.

[Den Hartog] Den Hartog J, Ten Kate T, Gerbrands JJ: Knowledge based interpretation of utility maps. CVGIP Image understanding, accepted for publication.

[Haralick] personal communication at ICDAR-93.

[Jonk]A.Jonk, A.Smeulders: An axiomatic approach to clustering line-segments. Submitted to ICDAR-95.

[Joseph] S.H. Joseph, T.P.Prodmore: Knowledge directed interpretation of mechanical engineering drawings. IEEE-PAMI 928-940, 1992.

[Kasturi] R.Kasturi et al.: A system for recognition and description of graphics. 9th IAPR Rome, 255-259, 1988.

[Koelma] D.Koelma, A.W.M.Smeulders: A visual programming interface for an image processing environment, PRL 1099-1109, 1994.

[Mayer] Mayer: Automatische wissenbasierte Extraktion von semantischer Information aus gescannter Karten, PhD thesis (in german) from Technische UniversitŁt M§nchen, 1994.

[Ten Kate] T.K. ten Kate: ROCKI - Raster to object conversion aided by knowledge based image processing. ESPRIT II 5376, Commission of the European Communities.

[Suzuki] S. Suzuki, T.Yamada: Maris: map recognition input system, PRL 919-933, 1990.

[Van Balen] R. Van Balen, T.ten Kate, D.Koelma, B.Mosterd, A.W.M.Smeulders: Scil-Image: A multi-layered environment for use and development of image processing software. In: Experimental Environments for Computer Vision and Image Processing, HI Christensen, JL Crowley (Eds.). World Scient Press, 1993, pp. 107 - 126.

Object-Process Based Segmentation and Recognition of ANSI and ISO Standard Dimensioning Texts

Dov Dori   Yelena Velkovitch   Liu Wenyin

Information Systems Engineering
Faculty of Industrial Engineering and Management
Technion, Israel Institute of Technology
Haifa 32000, Israel

Recognition of dimensioning text in engineering drawings is an essential part of the dimension understanding process, since this text is an important component of the dimension-set. We first introduce the OPD expression of the structural relations among the dimension-set components and the illustration of ISO and ANSI drafting standards. We then present principles and implementation of a method of segmentation and recognition of dimension text in ANSI and ISO based drawings. The method is vector-based, implying that the input is a set of vectors–wires (bars and arcs)–resulting from the orthogonal zig-zag vectorization, arc segmentation, and arrowhead pair recognition. Initial textbox extraction is done by a region growing process, performed on text-wire candidates. On the basis of textbox context (neighboring annotation wires), the drafting standard is detected. Raw textboxes are divided into logical textboxes, which are further decomposed into basic textboxes. A neural network based OCR algorithm is applied to each single character image decomposed from the basic textboxes and the symbol in each basic textbox in recognized. Finally, the OCR recognition results are verified by using contextual information and comparing the results with the measurements made directly on the drawing.

Keywords: text segmentation, text-graphics separation, text-graphics association, engineering drawing understanding, CAD conversion, OCR.

1. Introduction

Currently, engineering and industry make extensive use of Computer Aided Design for creating and processing of engineering drawings. However, by now, numerous paper-based engineering drawings have accumulated, and even if drawings are produced by CAD, frequently the only source is the paper hard copy. Since maintaining, modifying, and updating paper drawings manually is an expensive and time consuming process, it is desirable to have a system which will be able to understand paper based engineering drawings and translate them into CAD representation [1].

Our concern in this work is dimensioning text segmentation and recognition, which is an essential part of the drawing understanding process. This problem may roughly be separated into the following sub-problems: (1) text-graphics separation; (2) textbox extraction; (3) association of text with graphics; (4) text string recognition; and (5) recognition verification. While several studies have been conducted which are related to these problems, we are not aware of any research that has been directly targeted at this goal.

Fletcher and Kasturi [2] developed an algorithm for text string separation from mixed text/graphics images. It is based on the generation of connected components and the application of Hough Transform to group together the components into logical character strings which may then be separated from the text. Lai and Kasturi [3] presented a system for detecting dimension sets in engineering drawings that are drawn according to ANSI drafting standard. It is also based on the generation of connected components and their composition into text strings, which are associated with dimensioning lines. Neither one of these algorithms treats the problem of text/graphics connectivity nor do they detect text strings consisting of a single character.

Chai and [4] proposed an algorithm for textbox extraction, that is preceded by orthogonal zig-zag vectorization, arc segmentation, and arrowhead recognition. The textbox extraction is done by clustering the remaining short bars that are close to each other through a region growing process. As noted by the authors themselves, the algorithm is designed only for detection of text areas without string extraction, and may lead to some detection errors.

This paper describes a sub-system for dimension text recognition from engineering drawings produced according to either ANSI or ISO drafting standard, which is integrated into the Machine Drawing Understanding System (MDUS) [5]. The complete recognition is performed beginning with text segmentation and ending with verification of OCR results, including coarse segmentation (region growing), text/graphic connectivity handling, drawing standard determination, standard-dependent processing, character-recognition and verification.

We start by defining basic concepts of dimension sets and their components and explain the structural relations among them using the Object-Process Methodology (OPM) [6] in section 2. Expressing these structural relations using OPM helps analyze the segmentation and recognition process. Section 3 illustrates the differences between ANSI and ISO standards. In sections 4 and 5, the methods of text segmentation and text recognition are presented separately. Experimental results and summary follow in sections 6 and 7.

2. Dimension-set Components and Their Relations

In our work we define and use the following concepts and objects (see Figure 2.1). The structural relations among these concepts are expressed using the Object-Process Methodology [6].The definitions go from the basic level to the higher levels.

Bar, Arc and *Wire*: Bar is a straight line segment with non-zero width, Arc is a circular line segment with non-zero width, which can be thought as a bent bar, and Wire is a generalization of Bar and Arc. To express the structural relations among wires at a high level of abstraction, we use the term *cowiring* to denote the following relations between two wires: co-linearity if the two wires are bars, co-circularity with the same radius for two arcs, and tangency for a bar and an arc. Similarly, *orthowiring* is used to denote the orthogonal relations between two wires. Examples of wires and spatial relations between two wires are shown in Figure 1.

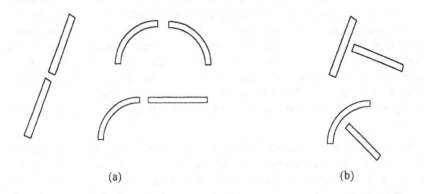

(a)   (b)

Figure 1: Spatial relations between two wires. (a) Cowiring. (b) Orthowiring (does not exist between two arcs).

Arrowhead, Tail and *Leader*: Arrowhead is a graphic object, whose shape is a solid isosceles triangle. We consider Arrowhead as a specialization of Bar, whose hard attributes (i.e., attributes that must be supplied to define the arrowhead) are Tip and Back, which are Bar's Endpoint1 and EndPoint2, respectively, and Width, which is the length of the triangle basis (i.e., the edge at the Back). Tail is a wire, which touches a cowiring Arrowhead at its Back to form an arrow, which we call *Leader*. Hence Leader is an aggregation of (i.e., it consists of) Arrowhead and Tail.

Leader pair is a pair of leaders, Leader1 and Leader2, that satisfy the structural relation "is-paired with", which abstracts two requirements: (1) that both Leaders are of the same specialization (i.e., both are bars or both are arcs) and (2) that their tails satisfy cowiring, which must be with zero gap in the case of ISO standard and some non-zero gap in the case of ANSI standard. The definition of this relation therefore provides a basis for a procedure to find Leader-pairs. As the OPD shows, Leader-pair is characterized by three attributes: Wiring, Pointing, and Continuity, each with two possible values (instances). Wiring has the values bar and arc, which is the type of wire of the two Tails of the Leaders in the pair. Pointing is the direction the two Arrowheads point to with respect to each other and its possible values are "inward" and "outward". Finally, Continuity is "positive" if the two tails touch each other or overlap, as in ISO standard, and "negative" if there is a gap between them, as in ANSI standard. The Cartesian product of these attribute values gives rise to $2^3 = 8$ possible Leader-pair type, which are shown at the bottom of Figure 2 along with their corresponding attributes values.

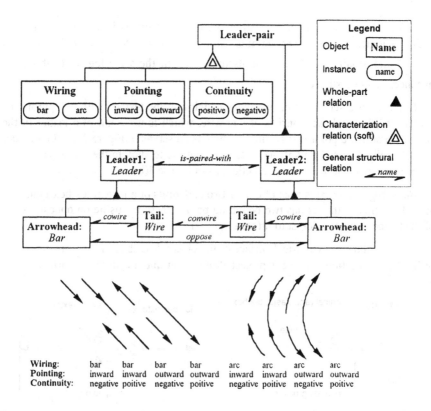

Figure 2: The structural relations among Leader-pair's components and the eight Leader-pair combinations.

Reference is a wire (bar or arc) at which a leader points. For a Leader to point at a reference, it must satisfy the requirement that the Arrowhead axis be perpendicular to the Reference and that the Arrowhead tip touches the Reference. This specification is sufficient to unambiguously define a procedure that determined for any pair of Leader and Reference whether the Leader points at the Reference.

Leader set is a single Leader or a Leader-pair.

Textbox is a minimal rectangle, enclosing text without any non-text element.

Basic textbox is a textbox, in which the text is a single string.

Logical textbox is a textbox, in which all elements are logically connected and refer to a common element of an engineering drawing. One logical textbox can contain from one to three basic textboxes. A logical textbox consisting of a single basic textbox may contain the nominal dimension alone or the nominal dimension with its tolerance (e.g., 5±1) or a range (e.g., R6–7). To incorporate tolerances, it may contain two basic textboxes (e.g., $25^{\pm 0.5}$), or three of them (e.g., $25^{+0.2}_{-0.3}$).

Sub-textbox is a textbox consisting of one or more basic textboxes enclosed within a logical textbox. The nominal dimension (e.g., 25 in the logical textbox $25^{+0.2}_{-0.3}$) and a pair of tolerance limits (e.g., $^{+0.2}_{-0.3}$ in the same logical textbox) are instances of sub-textboxes.

Guide set is a set of one or two wires which link the textbox with an appropriate leader set. The guide set consists of one wire, which is not a tail of a leader in the case of an outward pointing asymmetric dimension set, as in Figure 3(a). It coincides with the tail or pair of tails if the direction from the textbox to the arrowhead is the same as the arrowhead pointing direction, as in Figure 3(b).

Bounding points set is a set of one or two endpoints of a logical textbox guides or arrowhead tips. If the set has two bounding points, these points lie on opposite sides of the textbox and are closest to it (see Figure 3).

Site is a plane or cylindrical surface of the object described in the drawing. In an orthogonal projection, a plane is projected into a bar and a cylindrical surface – into an arc.

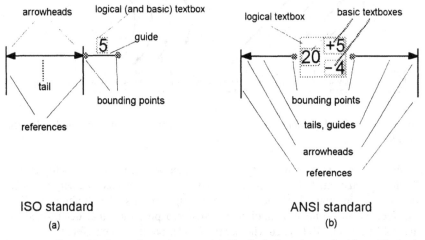

Figure 3: Example of two dimension sets. (a) The logical textbox coincides with the basic textbox; the textbox guide set is a single bar, which does not coincide with any one of the arrowhead tails. (b) The logical textbox consists of three basic textboxes; the textbox guide set coincides with the pair of tails, and the bounding points set is the pair of leader endpoints to the left and right of the logical textbox.

Dimension set is the top-level building block of dimensioning. It usually comprises Text and Leader-pair with the structural relation "is attached to", as shown in the OPD in Figure 4. Text consists of a Logical Textbox that contains Value, which expresses the Measure (distance or angle) between Reference1 and Reference2, which, in turn, are the attributes of Leader1 and Learder 2, respectively.

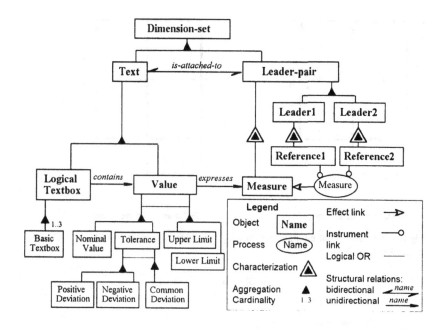

Figure 4: Leader-pairs and Text as the Dimension-set components.

3. ANSI and ISO Standards

The dimensioning annotation method is based on a detailed drafting standard, usually, ANSI or ISO. The standards state the requirements for the text location and orientation in the engineering drawing, as well as the size and relative position of the digits within the text. A typical engineering drawing produced according to ANSI standard is shown on the left hand side of Figure 5. The dimensioning text is located in a gap along the line connecting a pair of arrowheads, and it is always horizontal. A typical engineering drawing of ISO standard is shown on the right hand side of Figure 5, where the dimensioning text is positioned above or to the left (or right) of the leader–the line connecting a pair of arrowheads. Since ISO dimensioning text must be parallel to the leader, it may be tilted at almost any angle. Typical examples of the dimensioning text structure and its meaning are presented in Figure 6 and Figure 7 for ANSI and ISO drawings, respectively. ANSI standard allows for a larger variety of dimensioning text fonts than ISO standard (e.g., range values are not permissible by ISO standard).

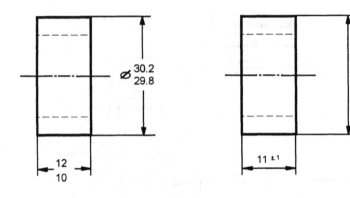

ANSI standard dimensions   ISO standard dimensions

Figure 5: Examples of engineering drawings produced according to ANSI (left) and ISO (right) drafting standards.

ANSI standard:

Nominal value:

20   50   (15)

(a)   (b)   (c)

Nominal value + tolerance value:

$20.0^{+0.5}_{-1.0}$ 34.2 ± 0.2 $15°15' \pm 20'$

(d)   (e)   (f)

Range values (minimal value + maximal value):

R5.9−7.1   R $\frac{7.1}{5.9}$ Ø $\frac{13.5}{12.8}$ 4XØ8.8−9.1

(g)   (h)   (i)   (j)

Figure 6: Examples of typical text from ANSI drawings.(a): Regular nominal value. (b): The text is enclosed within a rectangle. This means that the text describes a basic dimension [7]. (c): The text is enclosed in parentheses. This means that the text describes reference value. (d): nominal value is 20.0, plus tolerance is 0.5, minus tolerance is 1.0. (e): nominal value is 34.2, plus tolerance is 0.2, minus tolerance is 0.2. (f): nominal value is 15°15', plus tolerance is 20', minus tolerance is 20'. (g), (h): radius value lying in the range 5.9 to 7.1. (i): diameter value lying in the range 12.8 to 13.5. (j): 4 identical objects (e.g., holes) whose diameter is in the range 8.8 to 9.1.

ISO standard:

Nominal value:

20   (15)

(a)   (b)

Nominal value + tolerance value:

35.0 $-^{+\ 0.5}_{\ \ 1.0}$ 7.4 $-^{+\ 0.2}$ 7.4 $-^{+\ 0.2}$

(c)   (d)   (e)

Figure 7: Examples of typical text from ISO drawings.(a): Regular nominal value. (b): The text is enclosed in parentheses. This means that the text describes reference value. (c): nominal value is 35.0, plus tolerance is 0.5, minus tolerance is 1.0. (d), (e): nominal value is 7.4, plus tolerance is 0.2, minus tolerance is 0.2.

4. The Text Segmentation Method

The text segmentation algorithm consists of text-wire candidate selection, region growing, handling connectivity, standard determination and standard-dependent processing.

4.1 Text-wire Candidate Selection

Text-wire candidate is a vectorial wire (bar or arc) resulting from OZZ vectorization [5], which may potentially belong to text. The process of text-wire candidate selection is governed by the parameter *btext*, which determines the maximal character height in the drawing. Initially, all the wires are considered text-wire candidates. The elimination is done in two steps:

Step 1.

Eliminate all wires which satisfy at least one of the following conditions:

(1) The distance between the wire endpoints is greater than *btext*.

(2) The wire is a tail of a leader.

(3) The wire is a reference.

(4) Both wire endpoints coincide with tips of arrowheads comprising the leaders of a leader pair.

Step 2.

Eliminate each wire in a text-wire candidate chain (polyline) whose two endpoints coincide with endpoints of eliminated wires. All remaining wires are considered text-wire candidates.

4.2 The Region Growing Process

The region growing process is performed as described in [4], except that only the text-wire candidates are used as input to this process. Around each text-wire candidate, we construct a textbox as the minimal rectangle containing the wire. Before adding this new textbox to the textbox list, we examine all the textboxes already in the list and check whether any one of them is within a distance of btext at most from the candidate textbox. If such a textbox x is found, the candidate is merged with it, and the corners of x are updated to accommodate the added candidate. Another pass is then performed to check if, as a result of the last merger, two distinct textboxes can now be merged. The result of this procedure is a list of textbox candidates. From this list we delete the candidates which are too long and thin, and those whose area is too small.

Many of the textbox candidates obtained as a result of the region growing process do not exactly coincide with real textboxes due to one or more of the following reasons:

(1) Some textbox candidates may be false alarms.

(2) The detected textbox boundaries usually do not coincide exactly with the actual ones (see example in Figure 8(a)).

(3) The actual textboxes may enclose separate small angular dimension text symbols, such as °, ', and ", lying outside the textbox candidates. since no wire was found for these symbols during the vectorization process (see example in Figure 8(b)).

(4) Two or more logical textboxes may be merged into a single textbox.

(5) All textboxes are considered horizontal, whereas in fact some textboxes in ISO drawings may be vertical or tilted.

(6) In ANSI drawings, some dimensions may be enclosed within rectangles, denoting basic dimension in ANSI [7]. Such rectangles should be found and removed from the textboxes while keeping record of their existence.

(a)   (b)

Figure 8: Typical textbox candidates (denoted by grey rectangles) obtained as a result of the region growing process.

4.3 Handling the Text/Graphics Connectivity Problem

In order to separate connected text and graphics elements we should remove from the picture all pixels belonging to the graphic components, which may touch the text. Candidates for such graphic components are bars, arcs and arrowheads located close to textbox candidates detected by the region growing process. Detected boundaries of bars, arcs and arrowheads do not usually coincide with real ones, because of the following reasons: an object in the image is in general not an exact geometry object with straight boundaries; a wire may have variable width; there is noise along object boundaries; and algorithms for bar, arc and arrowhead detection have fixed accuracy. Therefore, in order to delete all pixels belonging to a graphic element, we remove all pixels within its expended boundaries. Expanded boundaries enlarge the detected graphic element. The expansion values depend on the quality of the drawing under consideration and the accuracy of primitive detection. In our system, the width of the wires is increased by a factor of 1.4, and their length is increased by two pixels at each endpoint. The boundaries of an arrowhead are expanded by two pixels in each direction.

Additional problems related to the separation of connected text/graphics elements may be encountered in ISO drawing and are considered in Section 4.5.

4.4 Standard Determination

We wish to determinate as early as possible the standard (which may be ISO or ANSI), according to which the drawing is produced, as this knowledge will be helpful for the subsequent processing. The standard determination is therefore done immediately after obtaining initial textbox candidates. The algorithm for standard determination is as follows.

The standard can be considered to be ANSI if there is a sufficient number (set by a predefined parameter, (see discussion below) of leader pairs which satisfy the following conditions (see Figure 9(a)):

(1) Each arrowhead in the leader pair is attached to a different bar-tail (i.e., C≠ D).

(2) Both these tails are either horizontal or vertical.

(3) The free end-points of each one (C and D) of these two tails are not connected by any common wire.

(4) The imaginary line CD connecting the closest endpoints of the tails (C and D) crosses a textbox candidate. dividing it into two approximately equal parts.

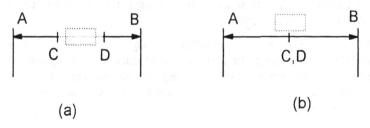

(a) (b)

Figure 9: Standard determination (Textboxes are denoted by grey rectangles).

The standard can be considered as ISO if there is a sufficient number of leader pairs which satisfy the following conditions (see Figure 9(b)):

(1) The tails of the arrowheads in the leader pair are identical, i.e., there is a continuous wire connecting the two arrowheads.

(2) This bar-tail is horizontal or vertical.

(3) A textbox candidate is located sufficiently close to the tail midpoint, and the tail does not cross this textbox candidate.

To determinate each standard, we define "sufficient number" of leader pairs which should satisfy the above conditions. This parameter depends on the quality of the drawing under consideration and the accuracy of the bar and leader-pair detection. In our system this parameter is set to be equal to 30% of the number of detected horizontal and vertical leader pairs for which a textbox candidate located close to the midpoint between the arrowheads has been detected.

At first glance, it might seem that some of the conditions for the standard determination are redundant. However, in our experimentation we have found that this redundancy is indeed necessary, as the above conditions cover different cases of relative positions of primitive and take into account various possible errors in primitives detection.

4.5 Standard-Dependent Processing

Having determined the standard we use special features of each standard to solve the problems associated with textbox candidates obtained as a result of the region growing.

ANSI Drawings Processing

We begin the processing of ANSI drawings with finding those textbox candidates which contain two or more logical textboxes and splitting them into single logical textboxes. There are two possible types of such textbox candidates, vertical and horizontal, as shown in Figure 10. The vertical type, shown in Figure 10(a), has vertical leader pairs, while the horizontal type has leader pairs whose projection on the horizontal axis is longer than their projection on the vertical axis, as shown in figures 10(b) and 10(c). We first split the textbox candidates belonging to the vertical type. Then we associate a unique leader set and bounding points set for each logical textbox candidate. We then separate the horizontal type of textbox candidate into logical textboxes, as discussed below.

Having separated textbox candidates into logical textboxes, we search for a rectangle (denoting basic dimension) around each logical textbox. Next, the boundaries of each logical textbox are adjusted to coincide with the real ones. Finally, basic textboxes are extracted from the logical textboxes. The details of each of these processes follow.

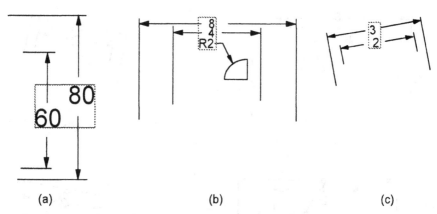

(a)   (b)   (c)

Figure 10: Examples of textbox candidates containing more than one logical textbox in ANSI drawings: (a) Vertical type; (b), (c): horizontal type.

Separating Textbox Candidates of Vertical Dimension sets

This procedure is based on the observation that the textboxes of adjacent dimension sets with vertical leaders are not collinear.

(a): Finding two intersecting white rectangles in opposite corners of the textbox candidate. (b): The two new textboxes obtained as a result of splitting the initial textbox candidate.

To extract logical textboxes from the textbox candidates of the vertical type, we search for intersecting white rectangles in opposite corners of the textbox candidate. If such rectangles are found, we separate the textbox candidate into two textboxes by splitting it along the sides of the white rectangles. This process is repeated for each one of the two resulting textboxes, until no more intersecting white rectangles are found.

Associating Textbox Candidates with Leader Sets

By considering leader sets and their neighborhood, we associate a unique textbox with each leader set. During this process we also find a guide set and a bounding points set for each textbox.

Separating Textbox Candidates of Horizontal Dimension Sets

We have associated a unique textbox with each leader set, but not necessarily a unique leader set to each textbox. We use the association information for separating the second type of the merged logical textboxes. If a textbox candidate is associated with more than one leader set, it means that it contains more than one logical textboxes. To separate these logical textboxes, we detect their horizontal axis – the line defined by the average of the Y-coordinates of the textbox bounding points (see Figure 11(a),(b)) – and split the corresponding textbox candidate along horizontal white runs that cross the textbox and are about equally spaced with respect to each axis (see Figure 11(c),(d)).

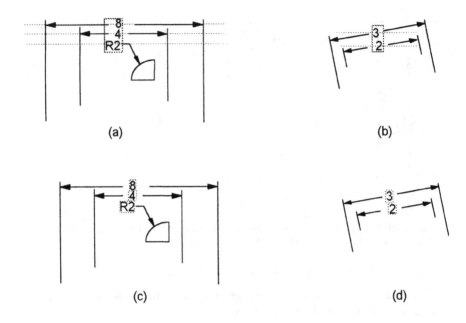

(a) (b)

(c) (d)

Figure 11: Examples of separation of merged logical textboxes of the second type in ANSI drawings. (a), (b): Finding axes of symmetry (denoted by dotted lines). (c), (d): Separated logical textboxes.

Finding Rectangles of Basic Dimensions and Textbox Boundaries Adjustment

For each logical textbox candidate we check for presence of a rectangle (denoting basic dimension) around its text. If such rectangle is found, it is removed and its detection is recorded.

Next, we adjust the boundaries of each logical textbox to coincide with the real ones by considering the pixel representation of the original engineering drawing image (we, also, take into account, that small symbols (such as °, ', and ")), may be initially outside the logical textbox candidate boundaries).

Basic Textbox Extraction

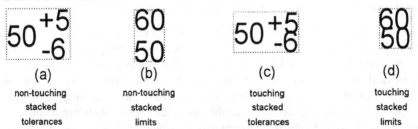

(a)	(b)	(c)	(d)
non-touching	non-touching	touching	touching
stacked	stacked	stacked	stacked
tolerances	limits	tolerances	limits

Figure 12: Examples of logical textboxes containing more than one basic textbox in ANSI drawings.

There are two possible structures of the logical textboxes containing more than one basic textboxes, as shown in figures 12(a) and 12(b). In most drawings, the basic textboxes are separated from each other by at least one white run as required by the standard. In spite of this, we may sometimes encounter cases of touching stacked characters as shown in figures 12(c) and 12(d).

Taking into consideration such cases, we first split logical textboxes containing non-touching basic textboxes. To do that, we search for white rectangles in the upper and lower left corners of the logical textbox. If the size of the white rectangles is within boundaries defined by parameters and there is a vertical white run along the whole logical textbox connecting the two white rectangles, then we extract a left and right sub-textboxes, and adjust their boundaries. If possible, we then divide the remaining logical textboxes and right sub-textboxes along a horizontal white run found in the their middle and adjust the textbox boundaries. Most of the left, upper and lower basic textboxes, are extracted by this procedure.

To separate touching stacked basic textboxes we first determine the typical character height. To this end, we construct the textbox heights histogram and set the typical text string height according to the mode (maximum occurrence) of the textbox heights. We then find the stacked limits logical textboxes and stacked tolerances right sub-textboxes as those whose height is close to twice the typical character height and divide them horizontally into two equal height basic textboxes.

ISO Drawings Processing

For ISO standard drawings we first process a special case of the text/graphics connectivity problem. Next, we find those textbox candidates which contain two or more logical textboxes and split them into single logical textboxes. We then associate each logical textbox with a leader pair and find a bounding points set for each logical textbox. The bounding points location is used to obtain the actual orientation of the textboxes, which, in ISO, is not necessarily horizontal. Next, the boundaries of each logical textbox are adjusted to coincide with the real ones. Finally, basic textboxes are extracted from the logical textboxes. The details of these processes follow.

Further Treatment of the Text/Graphics Connectivity Problem

While considering the text/graphics connectivity problem detected in section 4.3, we assumed that there are no "gross" errors in the bar detection and boundaries detected by the bar segmentation algorithm are rather close to real ones. However, in ISO drawings which contain touching text/graphics elements, the following bar detection error is possible. If text touches the tail of an arrow, as shown in Figure 13(a), two separate bars may be incorrectly detected instead of one continuous bar, as shown in Figure 13(b). Therefore we should reveal such errors and correct them. To perform this, for each leader pair we check the following conditions:

- There are two non-coinciding arrowhead tails in the leader pair.

- The two arrowheads in the pair point outward.

- There is a textbox close to the closest endpoints of the two arrowhead tails.

If all these conditions are satisfied for a leader pair, we connect its two tails into one continuous bar, which coincides with the real bar in the drawing, as shown in Figure 13(c).

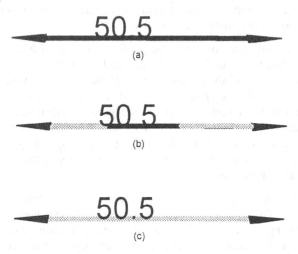

Figure 13: An example of a typical error in bar segmentation.

The detected bars are denoted by grey color. (a): Fragment of an original image. (b): Incorrectly detected bars. (c): The corrected results of bar detection.

Merged Logical Textbox Separation

To separate logical textboxes which are merged into a common textbox candidate, we split each textbox candidate along each horizontal and vertical bar crossing at least one of its sides and penetrating sufficiently deep into it (see Figure 14).

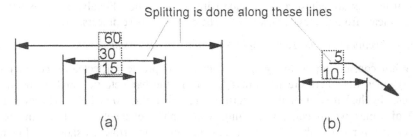

Figure 14: Separation merged logical textboxes in ISO drawings.

Associating Textboxes with Leader Sets

By considering leader sets and their neighborhood, we then associate a unique textbox with each leader set. During this process we also find a guide set and a bounding points set for each textbox.

Detection of Textbox Orientation

Each logical textbox is assigned a new coordinate system, obtained from the original one by rotating it counter-clockwise by an angle α, which depends on the position of the corresponding bounding points and the leader type. In the new coordinate system the logical textboxes are horizontal and the text is written from left to right rather than at any angle. Now, we construct these newly oriented textboxes in the original cooridinate system and adjust their boundaries to coincide with the actual ones.

Basic Textbox Extraction

We first divide the logical textbox into left and right sub-textboxes. This is done by considering the extent of variation the text top and bottom heights along the logical textbox. Then the right sub-textbox is divided into upper and lower tolerance basic textboxes as in ANSI drawings.

5. Text String Recognition

To perform text string recognition we divide the string into single characters, recognize each one of them separately, combine the whole string of recognized characters and improve the recognition result through geometric and numeric verification, as described below.

5.1 Single Character Segmentation

Usually, there is at least one white run separating the characters vertically from each other. However, sometimes the characters may touch one another. Therefore, for ANSI drawings, where all the basic textboxes are of the same size, we first divide all the text strings in the drawing into minimal rectangles, separated by at least one white run. Most of them are assumed to contain only one character, hence the histogram of the rectangle width has a mode, corresponding to the most frequent character width w. Having estimated w, we consider strings from right, upper, and lower basic textboxes and find rectangles having an approximate width of nw: (n = 2, 3, ...) and divide these rectangles into n equal parts. For ISO drawings, the character sizes within right basic textboxes may be smaller than those in left basic textboxes. We therefore generate the width histogram on the basis of right basic textboxes only, and use its mode for treating right basic textboxes as we do for ANSI drawings. Touching characters from left basic textboxes are not separated at this stage, because of the possible presence of the $\varnothing$ symbols, whose width may be about twice w.

Since touching characters significantly affect the recognized values, most of the touching characters, which are not detected by our algorithm, will be revealed during the verification process. This is true mainly for strings from left basic textboxes because they usually contain the dimension's nominal value, for which significant errors can be detected by our verification algorithm, described in Section 4.3).

5.2 Character recognition algorithm

23 characters (and symbols) may be encountered in dimension text: the ten digits (from 0 to 9) and the following 13 symbols: $\emptyset$, R, X, ~, (,), +, −, ±, ., °, ', ". To recognize them, we first identify each character as belonging to one of four classes according to the relative size of the minimal rectangle containing the character and its position in the string. The first class contains the ten digits and the eight symbols $\emptyset$, R, X, ~, (,), +, ±. The second class consists of the three small symbols, °, ', and " , used in angular dimension sets. Each of the third and fourth classes contains only one symbol, "−" and ".", respectively.

Having divided all the symbols into these four classes, we should recognize characters within the first two classes only, as the two other classes require no further processing.

To recognize characters within the first class, we first normalize the character size uniformly to a 12x12 image. This two-dimensional matrix is then converted into a one-dimensional array of 144 elements by raw stacking. This array is an input for a neural network [8]. We use a feedforward neural network consisting of three layers: the input layer has 144 neurons, which is the size of the input array; the hidden layer has 20 neurons (this number is obtained by trial and error based on our experiments); and the output layer has 18 neurons, which is the number of symbols in the first class.

We trained this neural network by a training set consisting of the following four parts: printed symbols of several fonts and sizes, symbols obtained from real life engineering drawings, symbols written by the author, and digits from a digit image data base [9]. The total size of the training set is 1585 samples. As expected, we found that the recognition accuracy increases as the training set size increases. We began our experiments with about half of the final training set size, and as the training set was increased up to the final size, the recognition accuracy grew from about 80 percent to 92 percent.

In a similar manner we perform the character recognition within the second class. Since the symbols here are smaller, the size of the normalized characters is 8x8 pixels and the structure of the network is as follows: input layer – 64 neurons, hidden layer – 5 neurons, and output layer – 3 neurons. The corresponding training set consists of 171 samples.

After single characters are recognized, they are concatenated into strings which are input the next process – text verification, described in the next section.

5.3 Numeric and Geometric Verification

The character recognition verification consists of the three steps: contextual verification within single basic textboxes, comparison between recognized dimension values and values measured from the drawing and verification of tolerance values according to relative dimension values. Each of these steps is described below.

Contextual Verification Within Single Basic Textboxes

We classify each basic textbox as belonging to one of eight classes according to its position within its logical textbox. Basic textboxes of the same class must have a common structure and satisfy the same requirements. To check if a basic textbox satisfies these requirements and to correct the OCR results, we use a bigram lookup table method and check several rules which cannot be expressed by the bigram table.

After the lookup table based verification is done, the rules which may not be expressed by the table are used. For example, one of the rules says: "If the first symbol is '0' and it is not followed by a period, it should be replaced by the symbol ' Ø' ".

Comparison Between Recognized Dimension Values and Values Measured from the drawing

First, we wish to determine the combined scale, i.e., the number of pixels per millimeter of the actual object described in the drawing (as opposed to resolution, which is the number of pixels per millimeter of the drawing). To do this, we construct a histogram of the ratio of the distance between arrowhead tips of the same arrowhead pair (measured from the drawing in pixel units) to the nominal value of the corresponding logical textbox. The ratio value corresponding to the histogram peak is chosen as the combined scale. Recognized nominal values, or range minimal and maximal values of the distance and angle are compared with the corresponding distances and angles, measured from the drawing. If they are not close enough, help of a human operator is called for.

Verification of Tolerance Values According to Relative Dimension Values

At this step we check each basic textbox containing tolerance values. Tolerance values are supposed to be significantly smaller than the corresponding nominal value. If this condition is not satisfied, a human operator help is used.

Note: This part of the verification process is not implemented in our system.

6. Experimental Results

22 engineering drawings containing 107 dimension sets and 253 symbols were processed according to our algorithm.

The recognition results for the main steps of the algorithm are summarized in Table 1.

The final accuracy of the logical textbox recognition, beginning with logical textbox segmentation and ending with contextual verification of the recognition results, is 80%.

Examples of segmentation and recognition results are shown in figures 15 and 16. All the dimensioning text in figures 15 was recognized correctly. In Figure 16, the segmentation was performed correctly, but there is an error in the text recognition within a string. The "125" string was recognized as "126". We expect that the first misrecognition will be corrected by the geometric verification (which is not implemented in our system), but the geometric verification cannot solve the second problem, because the error is in the least significant digit.

Table 1: Recognition accuracy for various steps of the algorithm.

Recognition Step	Recognition Rate Relative to Previous Stage	Absolute Recognition Rate
Logical textbox extraction	89%	89%
Dividing logical textboxes into basic textboxes	96%	85% (0.89 x 0.96)
Single character recognition before verification	92%	-
Single character recognition after contextual verification	95%	-
Whole text recognition within a logical textbox before verification	85%	72%
Whole text recognition within a logical textbox after contextual verification	94%	80% (0.85 x 0.94)

7. Summary

An algorithm for recognition of dimensioning text in engineering drawings was developed. Initial text regions are obtained by a region growing process, the drafting standard is determined and its features are used for text string extraction and association with the graphics. Next, a neural network based OCR algorithm is applied and its results are improved by applying a verification process.

All steps of the algorithm, except the geometrical verification and the tolerance verification, were implemented.

Further research should concentrate on both segmentation and recognition improvement. Text segmentation can be improved by using rules deduced after considering detection errors in a great number of real life engineering drawings. The OCR algorithm may also be improved by using a significantly larger set for training the neural network. Moreover, updating the neural network weights may be done automatically after each verification process, so that the system will in effect perform self-teaching. We also expect that geometrical verification and tolerance verification will significantly improve the recognition results.

References

1. D. Dori and K. Tombre, From Engineering Drawings to 3D CAD Models: Are We Ready Now? *Computer Aided Design,* **27**,4, pp. 243-254, 1995.

2. L.A. Fletcher and R. Kasturi, A Robust Algorithm for Textbox String Separation from Mixed Text/Graphics Images, *IEEE Trans. PAMI,* **10**,6, pp. 900-918, 1988.

3. C.P. Lay and R. Kasturi, Detection of Dimension Sets in Engineering Drawings, *IEEE Trans. PAMI,* **16**, 8, pp. 848-854, 1994.

4. I. Chai and D. Dori, Extraction of Text Boxes from Engineering Drawings, *Machine Vision Applications in Character Recognition and Industrial Inspection, Proceedings Series,* **1661**, SPIE, pp.38-49, 1992.

5. D. Dori, Y. Liang, J. Dowell and I. Chai, Sparse-pixel recognition of primitives in engineering drawings. *Machine Vision and Applications* **6**, pp. 69-82, 1993

6. D. Dori. Object-process Analysis: Maintaining the Balance Between System Structure and Behaviour. *J. Logic Computat.,* **5**. 2. pp. 227-249. 1995.

7. ANSI Y14.5M-1982 *−Dimensioning and Tolerancing,* The American Society of Mechanical Engineers. New York. 1983.

8. Paul Baffes, *NETS − neural network simulator.* Software Technology Branch NASA,   Jonson   Space   Center.   ftp   from   ftp.technion.ac.il /pub/unsupported/dos/simtel/neurlnet/nasanets.zip.

9. NITS database. ftp from host 129.6.61.25

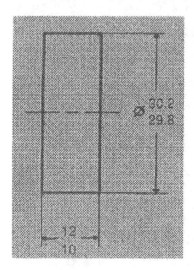 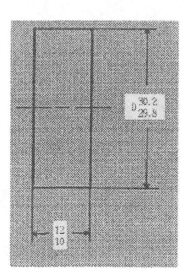

Figure 15: An example of dimensioning text recognition from the ANSI drawing. Left: The original image; Right: The results of text segmentation and recognition.

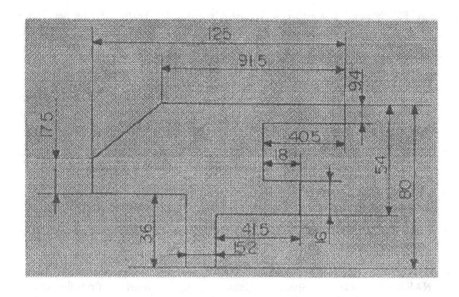

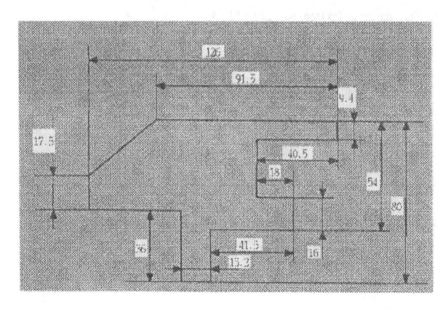

Figure 16: An example of dimensioning text recognition from the ISO drawing. Upper: The original image. Lower: The results of text segmentation and recognition.

A Combined High and Low Level Approach to Interpreting Scanned Engineering Drawings

Peter D. Thomas, Janet F. Poliakoff, Sabah M. Razzaq and Robert J. Whitrow

Department of Computing, The Nottingham Trent University,
Burton Street, Nottingham NG1 4BU, England.
E-mail: pdt@doc.ntu.ac.uk
Fax: +44 (0)115 948 6518

Abstract

The effective computer interpretation of engineering drawing remains a desirable aim yet it continues to provide academic challenge. Much early work was concerned with the interpretation of low level vectorised data. For simple drawings, direct association and interpretation of the low level data often provides a very effective technique but drawing data, whether linework or higher level textual information, can be subject to inaccuracies and uncertainties of interpretation. Thus drawing errors and problems introduced by scanning are likely to introduce ambiguities which cannot be resolved directly from the low level data. The approach described in this paper combines features of a low level approach based on node and vertex association with a higher level interpretation of the textual content of the drawing. The textual description of dimensions, etc. has previously been used by the authors and by others, for the correction of drawing structures, in some cases using 3-D reconstruction as a means of validating the data association. The present work attempts to model an aspect of human drawing interpretation, whereby an 'envelope of expectation' is developed, through the interpretation of dimensioning and annotation information. This approach allows a link to be established between the highest level information on the drawing (such as the title block) and the low level vectors of the three elevations. It is thus no longer necessary to interpret obscure detail within the vector data directly. Separation of text on the drawing using OCR techniques allows the field of interpretation for the linework to be significantly narrowed.

1 Introduction

An engineering drawing is a description of an object, which consists not only of geometric information but also of annotations and details of the manufacturing process. The geometric information is commonly contained in three orthographic projections, as shown in Figure 1. Each projection will normally include dimensioning and may contain additional textual information. The production of engineering drawings has changed greatly in recent years because of the use of Computer Aided Design (CAD) packages. The drawings produced by such packages are then be stored in computer data files for future updating and alteration. Furthermore, these computerised drawings may be used by the computer to generate 3-D reconstruction of the object, a task which was relatively difficult with the older paper drawings [1] - [4]. With increasing automation and integration of CAD with Computer Aided Manufacture (CAM) the use of such data files is growing in importance. Unfortunately many engineering drawing designs are available only on paper and cannot be used by a CAD system until they have been converted into a suitable format. Several groups [2] - [6] including our own are working on the automatic conversion of these paper drawings into computer readable form. The conversion process begins with the scanning and vectorisation of the drawings, then automatic interpretation takes place. Checking for consistency between the projections can be done and many errors can be corrected using redundancy and the dimensioning information in the projections. 3-D reconstruction allows the low level geometrical information in all three projections to be correlated and much work has been done in this area [5], [6].

Although low level processing is able to model some aspects of human drawing interpretation, another aspect of human interpretation has been largely overlooked. The engineer uses the textual information on the drawing to develop an 'envelope of expectation' and then goes on to identify the expected features. This accelerates the process of understanding the drawing. Even more importantly, the use of more complex information (dimensioning and annotation etc.) is vital to the successful interpretation, particularly in those cases where features are either omitted or distorted. For example, in the drawing in Figure 1, the phrase "6 HOLES DIA 5 EQUI-SPACED" leads an engineer to expect to find six holes each with a diameter of 5 and at an equal distance from each other, whereas only one is actually drawn. Thus, the engineer uses the textual information to locate the positions of the other five holes. The best which could be done by a "computerised interpreter" without this information would be to produce an object with only a single hole, as shown in Figure 2. In other cases, errors caused by scanning of the original drawing may lead to failure of the computer to interpret a drawing correctly. In Figure 1, the phrase "DIA. 10" allows the engineer to decide that the distorted shape is actually meant to be a hole and to replace it with a feature of the correct size and shape.

This paper builds on previous work by our group [5] on low level 3-D reconstruction and error correction using node and vertex association. We describe our progress towards implementation of the high level approach which involves extending the interpretation of textual information from simple dimensioning such as lengths and angles to include whole phrases such as "6 HOLES DIA 5 EQUI-SPACED". Yoshiura et al. [7] proposed a related method for drawing interpretation by entering complete grammatical sentences into the computer process. These sentences were derived by an engineer from the phrases on the original drawing and of the leader lines (arrowed lines) associated with the text. By contrast the aim of the work described here is to make use of the actual phrases found on the engineering drawing. These phrases occupy a linguistic domain greatly reduced in size but including abbreviations such as "DIA." which occurs with greater frequency in drawings than in normal usage. Thus, the range of possible phrases is significantly reduced because of the restriction to phrases in the engineering domain, with consequent implications for the speed and efficiency of the interpretation process. In what follows we describe the main steps in our combined high and low level approach and demonstrate its use in identifying and locating "missing" or distorted holes.

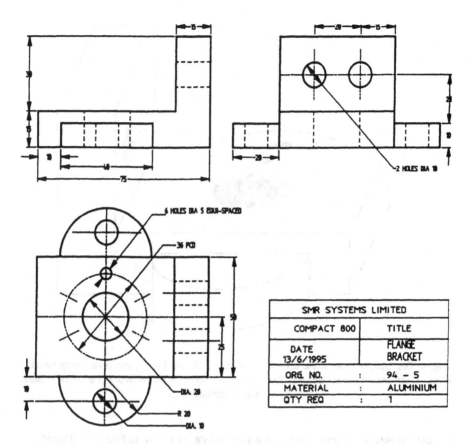

Fig. 1. An example of an engineering drawing with distorted and missing features. The text describes six holes but only one is actually drawn in the plan and none of them are shown in the other projections. In addition the hole in the front flange is distorted in the plan. This drawing cannot be interpreted correctly without using the information in the textual annotation. (The line types we discuss in this paper are listed in the Appendix.)

2 Low Level Interpretation

As explained above, the generation of a solid object by directly associating and interpreting the low level data from a simple engineering drawing is often an effective technique [8], [9]. Simple elements such as nodes and lines are combined to create vertices and edges, and then to construct surfaces. These elements are combined to create solid blocks, out of which objects are then assembled. However, errors in the drawing and distortions introduced by scanning are likely to introduce ambiguities not all of which can be resolved by direct interpretation of the low level data. In addition designers often simplify drawings and supply information in the form of

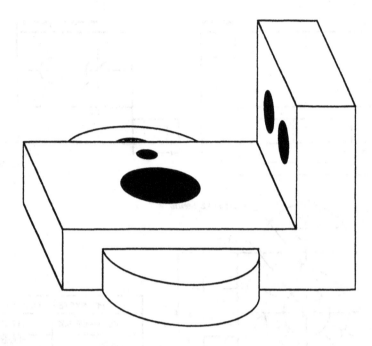

Fig. 2. The interpretation of the drawing in Figure 1 obtained using low level interpretation alone. Only one of the six holes in the top surface has been recognised and the hole nearest the front is missing.

textual annotation instead, where more appropriate, as shown in Figure 1. Therefore, an approach is needed that provides the information omitted by the simplifications and distortions. In order to generate a complete interpretation of an engineering drawing the information contained in the annotations on the drawing must be used.

3 The Combined High and Low Level Approach

The strategy for the automatic interpretation of interpreting engineering drawings using the high level details as well as low level data can be summarised as follows:

1 - The drawing is scanned and everything on the drawing (including text) is converted into a bitmap image.

2 - The captured bitmap image is vectorised (into short lines etc.) and, where appropriate, potential text blocks are identified for recognition using OCR.

3 - The low level vectorised linework is converted, wherever possible, to long line segments and circular arcs.

4 - The textual parts of the bitmap are analysed using OCR techniques to give alphanumeric phrases associated with particular positions on the drawing.

5 - The textual phrases are analysed using parsing techniques to generate associated syntactic nets.

6 - Meaning is generated using knowledge about engineering drawings which has been incorporated into the software.

7 - The meaning is used to provide extra data or to correct data in the vectorised data file (eg. the projections of the missing holes in Figure 1).

8 - The modified vectorised data file is then interpreted to reconstruct a 3-D model [5], [8], [9] using a combination of the information derived from both high and low levels.

Steps 1, 2 and 3 have already been described by other authors [1] - [9], while step 4 is accomplished by using OCR techniques. This paper concentrates on steps 5 - 7 where the high level information is analysed.

4 Engineering Drawing Text Interpretation

In general, interpreting text requires a vast quantity of knowledge. Both vocabulary and grammar rules are necessary for interpreting general textual information. However, in the case of the text on engineering drawings, the application domain can be much more limited. The vocabulary is restricted to technical terms and description of geometrical figures. Grammatical rules can also be restricted, because only assertive phrases need to be considered. The general process of analysing textual (phrasal) information on engineering drawings is expressed in the structural diagram in Figure 3. We have developed our algorithms initially for the interpretation of drawings containing three orthographic projections but clearly the principle could be extended to other types of drawings.

5 Our Approach

We have developed a strategy to derive meaning from phrases within the limited domain of engineering drawings. The phrases are assumed to provide information about sub-objects or dimensions of the object described in the drawing. (We allow sub-objects to include construction lines such as centre lines or pitch circles, although they are not "real".) For each phrase a Syntactic Net is generated, which carries information about the grammatical type of each component of the phrase and the relations between them. The text in each phrase is then interpreted and the information is used to generate a semantic template for the sub-object described. Before searching begins, the construction sub-objects are separated from the "real"

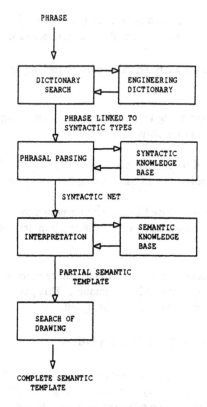

PHRASE

DICTIONARY
SEARCH

ENGINEERING
DICTIONARY

PHRASE LINKED TO
SYNTACTIC TYPES

PHRASAL PARSING

SYNTACTIC
KNOWLEDGE
BASE

SYNTACTIC NET

INTERPRETATION

SEMANTIC
KNOWLEDGE
BASE

PARTIAL SEMANTIC
TEMPLATE

SEARCH OF
DRAWING

COMPLETE SEMANTIC
TEMPLATE

Fig. 3. An overview of the process of parsing and interpretation of a phrase on an engineering drawing.

ones and processed first. This allows for cases where the identification of the "real" sub-object depends on construction sub-objects, such as in our example in Figure 1, where the six holes have centres lying on the pitch circle described by the phrase "36 PCD". The projections on the drawing are searched for matching features, which help to determine the actual location and other details of the specific sub-objects. The searching procedure uses the information in the template and the three projections in the drawing to find all the characteristic values required to recognise and display each object on the computerised drawing.

5.1. Dictionary Search and Phrasal Parsing

Our parsing strategy is to match a coded phrase to a syntactic net stored in the knowledge base. The syntactic net consists of a root node linked to a set of elements by specific relations, with, where possible, geometrical figures as the top ranking elements and, failing that, parameters or values. Each element can be a verbal syntactic type, such as an Object (or a Parameter or Adjective associated with an

Object), or it can be a numerical type, such as a Value or Number. These elements are defined in terms of restricted syntactic categories rather than general grammatical categories. The parsing process, therefore, leads more quickly to interpretations meaningful in the domain.

Initially, we identified the syntactic types which an engineering phrase is likely to contain and we use a single letter to denote each one. Most commonly, apart from simple dimensions, phrases refer to sub-objects as part of the whole object. All of the engineering drawings which we are considering represent three dimensional objects. Therefore we define "3-D Object" as an 'O' type and assign it as the highest ranking type in the syntactic network. The characteristics of the 3-D Object are then defined by the elements constituting our particular net. Other syntactic types which occur in relevant phrases include:- "Parameter", termed 'P', representing a sub-object's parameter such as diameter; "Adjective", termed 'A', which gives additional information about the sub-object; "Preposition" termed 'R', which describes the relation between two objects. In addition, the phrase may contain numbers which could refer either to a repeated sub-object (e.g. 6 HOLES) or to a value associated with the parameter of that object (e.g. DIA 5). These two terms are numerical syntactic types, either a "Number", 'N',where a number of objects are described, or a "Value", 'V', for a value associated with a parameter, depending on their positions in the phrase, and so no further processing is needed to determine their meaning. Thus, anticipating the analysis of our specific example, "6 HOLES DIA 5 EQUI-SPACED", the Syntactic Net contains Number (6), Object (HOLES), Parameter (DIA), Value (5) and Adjective (EQUI-SPACED). The corresponding Syntactic Net therefore contains the letters A, N, O, P and V, with 'A' linked to 'O' and 'V' linked to 'P', as shown in Net #3 in Figure 4. Thus, the relationships between the tokens or sub-phrases (usually words or numbers) are made explicit. This is especially important when processing more complex phrases which may involve several objects and many parameters and their values which must all be associated correctly with the objects. The sub-phrases play an important role in leading the parser through the phrase to an ultimate interpretation.

We have based our approach partially on the method of Chung et al. [10]. In order to interpret the textual information, we need to know which projections contain the sub-object described by the text, as well as how to pass the stored knowledge about the sub-object to the searching procedure.

The interpretation starts by pre-processing the phrase, i.e. splitting the phrase into separate tokens (sub-phrases, usually words or numbers) and matching each token in the phrase to an entry in the specially compiled engineering dictionary. The tokens in the dictionary are all assigned to a syntactic category. Thus, in our example phrase, "6 HOLES DIA 5 EQUI-SPACED", "HOLES" is identified as a sub-object which is classified in the Object category ('O') and "DIA" is recognised as a Parameter ('P') related to the Object. "EQUI-SPACED" describes additional information about the holes, and therefore is assigned to the Adjective category ('A').

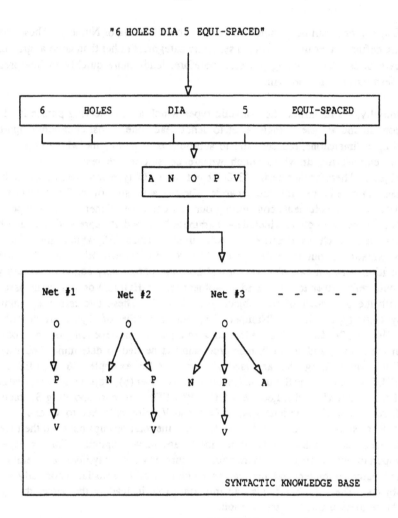

Fig. 4. An illustration of the linking of the phrase "6 HOLES DIA 5 EQUI-SPACED" to a syntactic net.

The numerical tokens are assigned to the Number ('N') or Value ('V') categories by a process which uses the position in the phrase (exploiting the fact that a value normally follows the parameter with which it is associated). Thus, in our example, the number of Holes (6) will be assigned to the 'N' category and the value of the parameter "DIA" (5) will be assigned to the 'V' category, as expected. The phrase has now been linked to the syntactic types 'N', 'O', 'P', 'V' and 'A'. The Syntactic Knowledge Base is then searched to find a matching syntactic net. The process of linking the phrase "6 HOLES DIA 5 EQUI-SPACED" to a syntactic net is illustrated in Figure 4, where it is linked to Net #3. In a similar way, the simpler example "2 HOLES DIA 10" yields syntactic types 'N', 'O', 'P', 'V' and is therefore linked to Net

#2. The phrase "36 PCD" yields initially only two tokens ("36" and "PCD") but "PCD" is further split into "PC" and "D". Thus, we have syntactic types 'V', 'O', 'P' and the phrase is linked to Net #1.

5.2. Interpretation and Searching

Once the phrase has been associated with a syntactic net, a Semantic Template is generated. The Semantic Template is associated with a particular sub-object (such as a simple hole) and specifies all the information required for that type of object (such as diameter, depth, position, orientation, etc.). In our example, the Hole Template will now indicate that there is a simple hole with diameter 5 and that this hole is repeated 6 times at equal spacing. Obviously, this information not sufficient by itself to identify the six holes on the drawing. Therefore, the procedure needs to know where these holes are located and it finds the location by searching the projections on the drawing. This searching is simplified by the fact that, in our system, phrases are stored together with a reference to the particular projection where the phrase was originally displayed, so allowing the search to be restricted initially to the projection where the phrase has appeared. The search is further simplified by exploiting information from the Leader Lines on the drawing. A Leader Line usually links a phrase to the geometrical object which is being described, as can be seen in Figure 1, so that the search can be limited initially to a small area as indicated by the leader line associated with the phrase. In the case of a hole the leader line points to the edge of the circular outline in a direction along a radius at that point. Thus, the searching function is not only focused on the appropriate projection and the most probable location on the drawing where it is likely to be displayed but it is also already armed by the knowledge base with information about the shape of the object. For example, the centre of a hole is usually indicated by the intersection point of two construction lines, eg. two centre lines or a pitch circle and a centre line. Such intersection points can be identified and tested to ascertain whether they represent the centre points of the holes and their coordinates can then be added to the template. A number of vectors approximating to a circle of the given radius confirms the presence of a hole. Figure 5 illustrates the process of associating a semantic template with the syntactic net for our example phrase. The search for the six holes in our example begins in the top view in the region at the end of the leader line and the first hole is identified when the crossing of the pitch circle with the centre line is found. The presence of the hole is confirmed by the vectors approximating to a circle of radius 5 centred on the crossing. The presence of the pitch circle together with the adjective "EQUI-SPACED" enables the other five centres to be found, even if some of the centre lines are missing.

5.3. Searching other projections

Once the sub-object has been located in one projection, the possible representations of the sub-object which may appear in the other projections are derived from the stored knowledge base. Thus, having located the projection of the sub-object, such

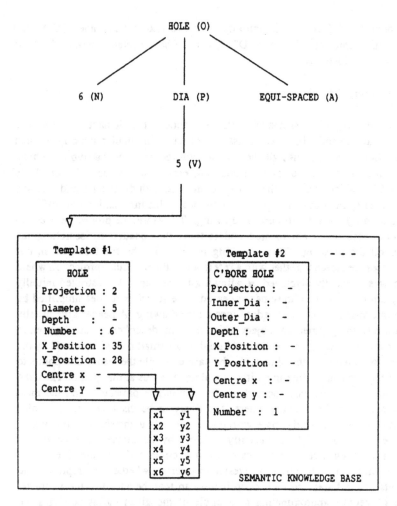

Fig. 5. An illustration of the association of a semantic template with the syntactic net linked to the phrase "6 HOLES DIA 5 EQUI-SPACED".

as the two circles representing the holes labelled "2 HOLES DIA 10" in Figure 1, the computer "expects" to find the remaining two representations of the hole depicted in the other two projections (which in this case are rectangles made up of both hidden lines and solid lines). The two projections are then searched for the appropriate representation of the sub-object; this search yields additional information about the sub-object, such as the position of the centre and, in this case, the depth of the hole (since holes are assumed to be penetrating unless specified otherwise in the text). Thus, additional information about the sub-object can be derived from the three projections to complete the semantic template. For the two holes the projections are found and the depth is found to be 15. Even when no representation is found in other

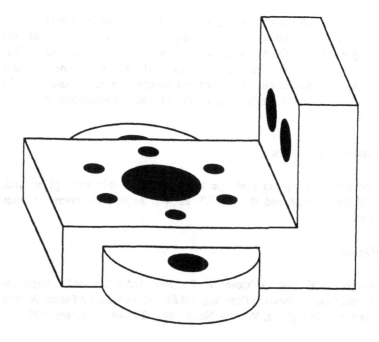

Fig. 6. The interpretation of the drawing in Figure 1 obtained using both high and low level interpretation. Comparing it with Figure 2, we see that all of the six holes in the top surface have now been recognised and the hole nearest the front has also been identified correctly.

projections, as, for example, in the case of the six holes, sufficient information can still be derived, such as the depth of the six holes, using the fact that the holes are penetrating (because the depth is not specified in the text).

6 Conclusions

This paper has demonstrated the validity of the combined high and low level approach. An algorithm is described for parsing textual information (on holes and their positions) in order to supplement the information from the low level interpretation. Without this knowledge from the textual information many holes would not be identified. In the case of the drawing from Figure 1, the interpretation of the phrase "6 HOLES DIA 5 EQUI-SPACED" enables all the six holes to be identified in spite of the fact that five of them are not shown in any of the outlines and one is shown in just one projection. The distorted hole is also identified correctly using interpretation of the phrase "DIA. 10". Figure 6 shows the resulting object complete with the set of six holes on the top surface and the hole in the front flange. The interpretation has been greatly simplified by making use of the fact that the domain is limited to engineering application. This method is an improvement on a previously reported method [7], which required an expert user to add sentences describing features on the drawing before interpretation could take place. Holes are one of the most common features described by the text on engineering drawings. Work is now in progress aimed towards extending the range of "knowledge" of our system and developing further algorithms for the interpretation of engineering drawings.

7 Acknowledgements

The authors gratefully acknowledge the support for this work given under the SERC/ACME scheme and the HEFCE as well as helpful comments from their colleagues.

8 References

[1] Joseph, S. H. and Pridmore, T. P., "Knowledge-Directed Interpretation of Mechanical Engineering Drawings", IEEE Transactions on Pattern Analysis and Machine Intelligence, Vol. 14, No. 9, pp. 928-940 September, 1992.

[2] Nagasamy, V. and Langrana, N. A., "Engineering Drawing Processing and Vectorisation System", Computer Vision Graphics and Image Processing, 49, pp. 379-397, 1990.

[3] Pavlidis, T., "A Vectoriser and Feature Extractor for Document Recognition", Computer Graphics and Image Processing, Vol. 35, pp. 111-127, 1986.

[4] Karima, M., Sadha, K. S. and McNeil, T. O., "From Paper Drawings to Computer-Aided Design", IEEE Computer Graphics and Applications, pp. 27-39, Feb., 1985.

[5] Poliakoff, J. F., Thomas, P. D., Razzaq, S. M., Shaw, N. G., "3-D Reconstruction for Correction of Errors and Imperfections in Scanned Engineering Drawings", COMPUGRAPHICS'95 (To be presented).

[6] Bergengruen, O., "About 3-D Reconstruction from Technical Drawings", International Workshop on Industrial Applications of Machine Intelligence and Vision, Tokyo, pp.46-49, April, 1989.

[7] Yoshiura H., Fujimura, K. and Kunii, T. L., "Top-Down Construction of 3-D Mechanical Object shapes from Engineering Drawings", IEEE Computer, pp. 32-40, Dec., 1984.

[8] Wesley, M. A., and Markowsky, G., "Fleshing Out Projections", IBM Journal of Research and Development, Vol. 25, No. 6, pp. 934-954, 1981.

[9] Sakurai, H., and Gossard, D. C., "Solid Model Input Through Orthographic Views", ACM Computer Graphics, Vol. 17, No. 3, July 1983.

[10] M. Chung, M., and Moldovan, D., IEEE Expert, pp.36-44, 1994.

Appendix

We show in Figure A1 below the different line types discussed in this paper.

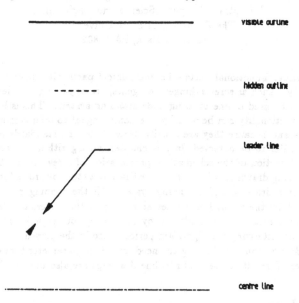

Fig. A1. Examples taken from Figure 1 to illustrate each of the different line types discussed in this paper.

The line types are:

Visible Ouline - part of the outline of the object which is visible in the given projection.

Hidden Ouline - part of the outline of the object which is not actually visible in the given projection, because it is behind another part of the object.

Leader Line - line used to associate a piece of text with the part of the object to which it refers, for example a hole.

Centre Line - a line used to indicate a centre of a feature on the drawing in a particular projection, for example of a hole.

Functional Parts Detection in Engineering Drawings: Looking for the Screws

María A. Capellades[1] and Octavia I. Camps[1,2]

[1] Dept. of Electrical Engineering
[2] Dept. of Computer Science and Engineering
The Pennsylvania State University
University Park, PA 16802

Abstract. Functional parts – i.e. mechanical parts with intrinsic functionality – such as screws, hinges and gears, are appealing high level entities to be used in line drawing understanding systems. This is because their functionality can be used by a reasoning agent to infer surrounding objects and because they are usually drawn following standards making them easier to be detected. In this chapter, an algorithm for the automatic detection of the schematic representation of screws in mechanical engineering drawings is being presented as a first step towards a function-based line drawing understanding system. All the running parameters required by the algorithm are set according to the American National Standards Institute standards and by using a rigorous experimental protocol characterizing the algorithm performance in the presence of image degradation, thus eliminating the need for *ad hoc* parameter tuning. Experimental results on several real line drawings are also presented.

1 Introduction

The automatic interpretation of engineering drawings remains a difficult task due to their complexity and diversity. In particular, the dual nature of engineering drawings, carrying information in both graphical and textual form, coupled together with the fact that images of paper drawings are, in general, of poor quality and vary with the different individual's drawing styles add more difficulty to the analysis.

We propose to bridge the gap currently existing between the early segmentation stage (lexical stage) and the final interpretation stage (semantic stage) of the interpretation process by using *functional models* as high-level reasoning entities and *physics-based degradation models* to eliminate *ad hoc* tuning of parameters.

Most man-made objects are made to perform one or more functions, and these functions dictate their shape. The concept of function-based vision is not new. It can be found in the literature since the late 70's [15, 19, 2, 3, 6, 14] and has received significantly more attention [7] after the work by Stark and Bowyer [17]. However, until now, line drawings understanding systems have not incorporated any of these advances.

The use of *function reasoning* in line drawing understanding will not only ease the interpretation task, but it will also help to create more complete models with functional information. Thus, we use as high-level reasoning entities mechanical parts with intrinsic functionality. We call these parts, *functional parts*. Examples are screws (tie parts together), hinges (provide articulations between parts), gears (convert power ratios), etc. Functional parts are particularly appealing since:

1. their functionality and inter-relationships with surrounding parts can be used by a high-level reasoning agent to infer the remaining objects in the drawing;
2. they are usually drawn following standard representations, thus they are potentially easier to identify using pattern recognition techniques than arbitrary parts;
3. they provide logical units that can be used to query a database of drawings (consider for example, the task of recalling all parts that have a given type of screw, known to be defective).

However, before a system can reason on any high-level entity, it must first segment the drawing to extract these entities. The results obtained at this first stage, on which all following stages rely on, are highly dependent on the image quality of the document. This quality is degraded by processes commonly used on documents such as printing, photocopying, and scanning. Thus, it is important to take these problems into account when designing the feature extraction algorithms. This is usually done in an *ad hoc* manner by performing expensive trial and error runs to tune the running parameters of the algorithms. Recently, Kanungo *et al* [9] have proposed a physics-based model for the local distortions introduced by printing and scanning processes. We use this model to generate degraded *synthetic* data to characterize the performance of the detection algorithms and to *automatically* select their optimal running parameters.

In this chapter, we focus in developing a tool for the recognition of screws in mechanical engineering drawings following the American National Standards Institute (ANSI) standards for schematic thread representation [13], as a first step towards a function-based line drawing understanding system. The input to the algorithm consists of a list of all the lines found using a modified version of the Orthogonal Zig-Zag (OZZ) algorithm [4]. The screws are detected using knowledge about their standard representation (shown in Fig. 1). All the running parameters required by the algorithm are set according to the ANSI standards and by using a rigorous experimental protocol characterizing the algorithm performance in the presence of image degradation.

2 Screw-Detection Algorithm

The screw-detection algorithm takes advantage of the unique structure of the schematic representation of threads in section in standard drawings [13]. In this representation (see Fig. 1), the threads are drawn as parallel lines separated all by a fixed distance, and with a periodicity in width and length. The difference in

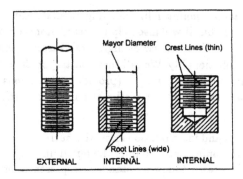

Fig. 1. Schematic Representation of Threads in Section

the lengths of two consecutive lines cannot exceed a given value, and the same applies for the distance between them. All these features must be taken into account when looking for parallel lines that match this pattern. If the number of the consecutive lines that meet the requirements specified above is larger than a threshold, then a screw is said to be found.

2.1 Line Detection: The OZZ Algorithm

The first step towards detecting the screws is to find all the lines in the drawing. This task can be accomplished by using a modified version of the Orthogonal Zig-Zag (OZZ) algorithm [4] briefly described next.

The OZZ algorithm for line detection was developed by Dori *et al.*[4]. The main reason why we use this algorithm is because it is sparse pixel: it avoids massive pixel addressing, visiting some of the pixels in the image only once and never visiting the rest of them. The image is scanned horizontally and then vertically skipping a number of pixels between one scanning cycle and the next. The number of pixels to be skipped is a parameter called *screen-skip*, and its value is determined depending on the minimum line length to be recognized in the drawing.

The implementation of the OZZ algorithm used in this work differs in certain aspects from the original one described in [4]. One of the problems with the OZZ, and with almost any line detection algorithm, is that under the presence of noise it may break a line into shorter segments. In [4], a series of tests are performed to merge line segments. This is done for every possible pair of lines, until no more merging can be done. This results in a computational complexity quadratic in the total number of lines N, that can be devastating if the drawing in consideration is fairly complex. Instead, we have developed a new merging algorithm with complexity proportional to $N \log N$, described next.

2.2 Merging Multiple-Segment Detected Lines

As it was mentioned before, some lines will be broken by the OZZ algorithm into several shorter segments due to noise and other artifacts. In order to correct this, a merging procedure was developed that does not require to check every possible pair of lines. The procedure takes advantage of the fact that the several segments in which a line breaks into are approximately collinear and have more or less the same width.

The procedure is slightly different depending on the direction of the lines to be merged. Say that we are interested in merging horizontal segments. The list of horizontal segments detected by the OZZ is stored in a heap and sorted by their row coordinates using a heapsort [16] with complexity $N \log N$. A checking procedure is then performed for every group of segments that have row coordinates differing by no more than a threshold *maxdispersion*. This parameter accounts for the maximum shifting in the row coordinates of the multiple segments into which a line was broken by the OZZ. Then, the algorithm goes through every group of segments that have the row coordinates within *maxdispersion*, sorts the segments by the column coordinates of one of their endpoints, also using a heapsort, and checks if the column coordinates of opposite endpoints of consecutive segments in the list differ within a tolerance. If they do, then the two segments are merged, and the checking continues. This is illustrated in Fig. 2a, where checking is performed only in the small square regions between segments.

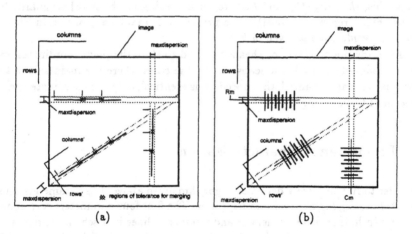

(a)   (b)

Fig. 2. (a) Merging multiple-segments procedure (b) Screw-detection procedure

For the vertical lines the procedure is identical, only interchanging rows and columns. The process is slightly more complicated for slanted lines, although the philosophy is the same. In this case, one more step is needed before the procedure described above can be applied: the slopes of all the slanted lines found in the drawing need to be sorted using a heapsort. Then, the lines are grouped according to their slope and the previous procedure is applied to each of these groups (Fig. 2a).

2.3 Detecting Horizontal Screws

We describe the algorithm for finding screws with horizontal axes. The algorithms for the other directions are similar. Potential screws with their main axes in the horizontal direction have their root and crest lines vertical. The algorithm first finds groups of vertical lines that have the row coordinate of their midpoints, R_m, within *maxdispersion*. For every one of these groups, the segments are then sorted by the column coordinates of one of their endpoints (both endpoints have the same value for the column coordinate in this case). The algorithm then goes through this sorted list, and for every two consecutive segments in the list three tests are performed. If the width of the segments is less than a threshold th_w, then the algorithm checks if the distance between the two segments is less than a threshold th_d. The third test consists on checking if the difference in the lengths of the two segments is less than a threshold th_l. If the three tests are passed, then a counter is incremented and the checking continues; if not, then it checks if the counter value is larger than th_n. In the case that the counter is greater than th_n, then the hypothesis that a screw has been found is validated, the coordinates of the starting endpoint of the first line meeting the requirements and the ending endpoint of the last line are stored in a file, and the search continues until no more segments in the list of vertical lines are left. All the thresholds, th_w, th_d, th_l, and th_n are set according to the ANSI standards for screw thread representations[1]. The procedure is graphically shown in Fig. 2b and the pseudo-code is given in Fig. 3.

As in the previous section, heaps are used to store the point coordinates and the sort algorithm used is a heapsort. For the cases where the main axis of the screw is vertical or slanted, the procedure is identical, only the coordinates being used need to be changed.

3 Performance Characterization

In order to run the screw detection algorithm given above, five parameters must be set: th_w, th_d, th_l, th_n and *maxdispersion*. The first four parameters correspond to the width, interdistance, length and number of lines in a schematic representation of a screw, respectively. Thus, they are easily set by following the ANSI standards. The fifth parameter *maxdispersion* is the tolerance for the misalignment of the midpoints of the lines representing the screw. Since the alignment of these points is severely affected by the image quality, setting this parameter to a suitable value requires more care.

In the sequel, an experimental protocol for the performance characterization of the algorithm when the image is increasingly degraded and the parameter *maxdispersion* is varied is presented. This characterization not only shows the validity of the algorithm by illustrating how it behaves in the presence of image degradation, but also provides a tool to select the optimal value of the parameter *maxdispersion* to minimize the probabilities of misdetection and false alarm.

SCREW DETECTION ALGORITHM

```
For every vertical line in the image
    do
    Store its midpoint row R_m in a heap
    end do
Sort the heap
Group the lines with R_m within maxdispersion
For every group of lines
    do
    Store the column of one endpoint in a heap
    Sort the heap
    Initialize search
    For every two consecutives lines in the heap
        do
        If their widths < th_w
            and
            the distance between them < th_d
            and
            the difference in their lengths < th_l
        then
                increment counter
        else
                end search
        If the search ended
        then
            Initialize search
            If counter is > th_n
                then
                store results in output file
    until no more lines are left in the heap
```

Fig. 3. Pseudo-code of the detection algorithm for horizontal screws.

3.1 Experimental Protocol

The performance characterization of the algorithm was done following the guide-lines given in [5] and [10]. In particular, the methodology in [10] is very useful because it allows to integrate a large number of operating curves relating the probabilities of mis-detection and false alarms for each parameter setting into a single performance curve.

The methodology consists of two steps of standard decision analysis and two steps inspired by psycophysical methods [10].

Step 1: First, the two noise-free images shown in Fig. 4a and Fig. 4b were generated. Fig. 4a is a *target* image consisting of an ideal screw drawing and it will be used to estimate the probability of a misdetection. Fig. 4b is a *no-target* image consisting of a drawing closely resembling the one shown in (a) but that does not follow the ANSI standards. This image was designed to estimate the probability of false alarm.

The images were then degraded to simulate an increasing number of duplication using the perturbation model given in [9]. According to this model, foreground and background pixels are changed following exponential distributions. The probability of a foreground pixel changing to the background was set to $P(0 \mid d, f) = \exp(-0.9d^2)$ and the probability of a background pixel changing to the foreground was set to $P(1 \mid d, b) = \exp(-2d^2)$ where d is the inverse distance, f is foreground and b is background. A total of 1000 perturbed images were generated, divided in five sets corresponding to five different levels of perturbation. The level of perturbation was varied by changing the threshold value th_b (th_f) used to decide if a background (foreground) pixel is actually changed given that its probability of changing indicates it should. The values used for these thresholds were: level 1, $th_b=th_f=0.9$; level 2, $th_b=0.8$, $th_f=0.9$; level 3, $th_b=0.8$, $th_f=0.7$; level 4, $th_b=0.8$, $th_f=0.5$; and level 5, $th_b=0.8$, $th_f=0.3$. Images for levels 1 and 5 are shown in Fig. 4. Note that as the perturbation level is increased the target and no target images become more and more similar.

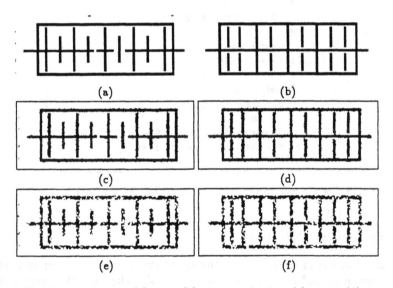

Fig. 4. Test images, level 0: (a)target (b)no-target, level 1: (c) target (d) no-target, and level 5: (e)target (f)no-target

Step 2: The detection algorithm was run on the test images for all the perturbation levels and for values of the parameter *maxdispersion* equal to 0.01, 0.02, 0.04, 0.06 and 0.08 inches. Operating curves of the probability $P(misdetection) = P(no-target \mid target)$ plotted against the probability $P(falsealarm) = P(target \mid$

no − target) for each of the perturbation levels and *maxdispersion* values were generated.

Step 3: From the operating curves, the value of the probability of error $P(E)$ was calculated for each level and each value of *maxdispersion*. The probability of error was taken as the average of the probabilities of false alarm and misdetection when the last one is equal to 0.2.

Step 4: A plot of $P(E)$ versus level of perturbation is then obtained for each value of *maxdispersion* (Fig. 5a). From this plot, the *critical signal variable* defined as the maximum level of perturbation that an image can have in order to get a probability of error $P(E)$ less than 0.16 is measured. Finally, a plot of the critical signal variable versus the variable of interest, *maxdispersion*, is obtained (Fig. 5b).

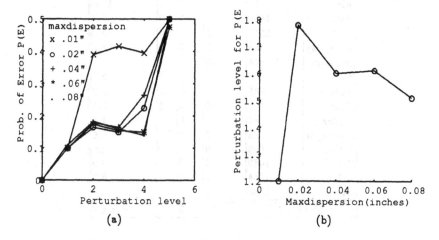

Fig. 5. Results of the Performance Characterization.

From the results shown in Fig.5b, it is seen that the optimal value for the *maxdispersion* parameter − i.e. the *maxdispersion* value corresponding to the maximum level of perturbation and $P(E) = 0.16$ − is between 0.02 and 0.04 inches. In practice it is very difficult to find images degraded more than a level 2. Usually, we find images in the range between 1 and 2.

4 Results with Real Images

The screw-detection algorithm was also tested on fourteen scanned images of modified real mechanical engineering drawings from [12] with up to 10 screws. Some of the images were scanned at a resolution of 300 dpi while others with smaller details were scanned at 450 dpi. All of them were globally thresholded.

The running parameters for the algorithm were set as: *maxdispersion*=0.03", th_w=0.025", th_d=0.15", th_l=0.15", and th_n=5. Table 1 lists for each of the test images, the number of line segments output by the OZZ algorithm, the number of screws in the drawing, the number of misdetections, the number of false alarms, and the running time on a SUN SparcStation 5. Although the number of real images is low since they are difficult to obtain, the overall misdetection and false alarm rates are also given. These rates were 17.6% and 9.8%, respectively. However, three of the false alarms in two of the tested images were found among the text in the drawing. Thus, if the algorithm were to be run after segmenting out the text, the false alarm rate would go down to 3.9%.

Table 1. Results on Real Images

Image	Segments	Screws	MD	FA	Time (sec.)
Air1	2311	9	3	1	10.3
Air2	1607	7	1	0	7.1
Air3	1243	1	0	0	3.6
Assembly	798	2	0	0	3.2
Bearings	1548	10	3	1	5.9
Bolt	256	1	0	0	2.2
Bplate	641	2	0	0	3.2
Cast	870	0	0	2	3.9
Collar	141	2	0	0	1.4
Conveyor	978	2	0	0	2.6
Holes	508	0	0	0	2.0
Joint	1069	4	0	0	3.0
Lift1	1344	3	1	1	5.2
Stand	1380	8	1	0	4.7
TOTAL		51	9	5	Aveg. 4.2

Three of the test images, their segmentation using OZZ and the bounding boxes of the detected screws are shown in Figs. 6-8. Fig. 6 is an example of a drawing where all screws were detected and there were no false alarms. Fig. 7 is an example where there was one misdetection because the OZZ algorithm detected only four segments for the bottom left screw. Finally, Fig. 8 is an example with no screws that resulted in two false alarms. Interestingly enough, the false alarms are not located on the detail representation of the screw but on the text area, and could have been avoided by segmenting out the text first.

5 Implementation Details

The algorithms were implemented using the C language [11]. The input to the program is a bitmap image in PBM format, or *portable bitmap* format. There are three different modules in the program, one for each one of the three main groups of lines: vertical, horizontal and slanted lines. Each module performs the OZZ algorithm to detect the corresponding lines, merges the segments and detects the presence of schematic representations of screws. At the end of each module, the information extracted is stored in an output file. For lines, the information consists of the endpoints and width of the line. For the screws, the information consists of the coordinates of the upper-left corner and bottom-right corner of the bounding box of the screw.

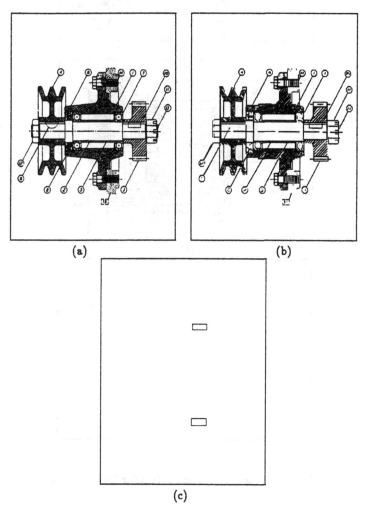

(a)   (b)

(c)

Fig. 6. *Assembly* image: (a)Original (b)after OZZ (c)screws

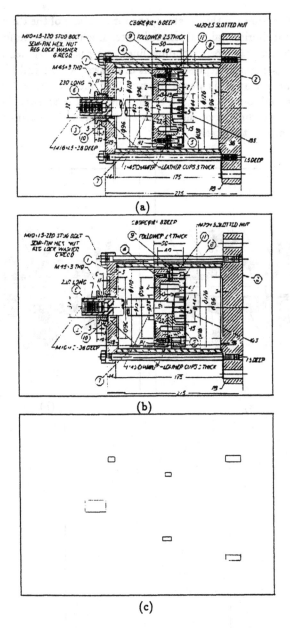

Fig. 7. *Air2* image: (a)Original (b)after OZZ (c)screws

257

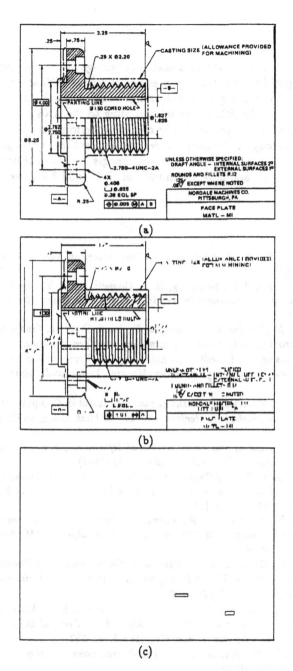

Fig. 8. *Cast* image: (a)Original (b)after OZZ (c)screws

6 Summary

Functional parts can be extracted taking advantage of their standard representation. An algorithm to detect the presence of screws with schematic representations in mechanical engineering drawings was presented. All the parameters needed to run the algorithm were set according to the ANSI standard or using a rigorous experimental protocol. The algorithm performance was characterized by running it through 1000 images under five different levels of degradation. The algorithm was also tested on fourteen real images of moderate complexity with overall misdetection and false alarm rates of 17.6% and 3.9%, respectively.

References

1. ANSI Y14.6: Screw Thread Representation. *American National Institute of Standards*, (1983)
2. Brady, M., Agre, P. E., Braunegg, D. J., Connell, J. H.: The mechanics mate. In T. O'Shea, editor, *Advances in Artificial Intelligence*, pages 79–94. Elsevier, New York (1985)
3. Di Manzo, M., Trucco, E., Giunchiglia, F., Ricci, F.: FUR: Understanding FUnctional Reasoning. *International Journal on Intelligent Systems*, 4 (1989) 159–183
4. Dori, D., Liang, Y., Dowell, J., Chai, I.: Sparse–Pixel Recognition of Primitives in Mechanical Engineering Drawings. *Machine Vision and Applications*, 6 (1993) 69–82
5. Haralick, R.: Performance Characterization Protocol in Computer Vision. *Proc. of NFS/ARPA Workshop on Performance versus Methodology in Computer Vision*, June (1994)
6. Ho, S.: *Representing and using functional definitions for visual recognition*. PhD thesis, University of Wisconsin, Madison, WI (1987)
7. IEEE Computer Society Technical Committee on Pattern Analysis and Machine Intelligence. IEEE Computer Society Workshop on the Role of Functionality in Object Recognition. june (1994).
8. Joseph, S. H., Pridmore, T. P.: Knowledge–Directed Interpretation of Mechanical Engineering Drawings. *IEEE Trans. on Pattern Analysis and Machine Intelligence*, 14(9) (1992) 928–940
9. Kanungo, T., Haralick, R. M., Phillips, I.: Global and Local Document Degradation Models. *Proc. of Second International Conference on Document Analysis and Recognition*, October (1993) 20-22
10. Kanungo, T., Jaisimha, M. Y., Palmer, J., Haralick, R.: A Methodology for Analyzing the Performance of Detection Algorithms. *Proc. of the 4th International Conference on Computer Vision*, May (1993) 247–252
11. Kernighan, B. W., Ritchie, D. M.: *The C Programming Language*. Prentice Hall, Murray Hill, NJ (1988)
12. Luzadder, W. J.: *Fundamentals of Engineering Drawing*. 8th edition, Prentice Hall, Englewood Cliffs, NJ (1981)
13. Luzadder, W. J., Duff, J. M.: *Fundamentals of Engineering Drawing*. Prentice Hall, Englewood Cliffs, NJ (1993)
14. Minsky, M.: *The Society of Mind*. Simon and Schuster, New York (1985)

15. Rosch, E., Mervis, C. B., Gray, W. D., Johnson, D., Boyes-Braem, P.: Basic objects in natural categories. *Cognitive Psychology*, **8** (1976) 382–439

16. Sedgewick, R.: *Algorithms*. Addison–Wesley Publishing Co., New York (1988)

17. Stark, L., Bowyer, K.: Achieving generalized object recognition through reasoning about association of function to structure. *IEEE Transactions on Pattern Analysis and Machine Intelligence*, October (1991) 1097–1104

18. Vaxiviére, P., Tombre, K.: CELESSTIN: A System for Conversion of Mechanical Engineering Drawings into CAD format. *IEEE Computer: Special Issue on Document Image Analysis Systems*, July (1992) 46–54

19. Winston, P. H: Learning structural descriptions from examples. In P. H. Winston, editor, *The Psychology of Computer Vision*. McGraw-Hill, New York (1975)

Reconstruction of 3D Solid Model from Three Orthographic Views – Top-Down Approach

Ken Tomiyama* and Ken'ichi Nakaniwa

Department of Mechanical Engineering
Aoyama Gakuin University
6-16-1 Chitosedai, Setagaya, Tokyo 157 JAPAN
Voice: +81-3-5384-1111, Fax: +81-3-5384-6314
E-mail: tomiyama@alice.me.aoyama.ac.jp

Abstract. A method of reconstructing 3D solids from three orthographic views is presented here. The method utilizes a "top-down" approach where a series of intermediate 3D solids that contain the target object are constructed and carved into the target object. Consequently, operations must be in 3D and are mostly subtraction operations. Three views of the intermediate 3D solids are generated and compared with the original three views to determine the needed sequence of 3D operations. This provides a natural feedback mechanism to the proposed method. The method was applied to forty five samples with twenty one successes and four near successes. Discussions include subjects for further improvements.

1 Introduction

Engineering drawings in the form of a set of three orthographic views have long been in existence. Design engineers have been trained to understand drawings that contain various markers with special meanings as well as to visualize three dimensional (3D) images of indicated objects. The method of drawing has developed from hand-drawing on sheets of paper to computer generation using 2D CAD software packages. Recently, 2D CAD software packages have been gradually replaced by more sophisticated 3D packages where object design is performed in a 3D environment.

On the other hand, there still are many hand-drawn paper-based drawings and 2D CAD data. There is a need, therefore, to convert paper drawings to 2D CAD and then to 3D CAD data. There are many studies on this conversions, some of which are listed in the Reference section ([15] [5] [11] [7] [6] [14] [8]). The initial phase of our study is focused on this technique of automatic conversion of hand-drawn paper-based drawings to 2D CAD data [9] [10].

The effort is then shifted to concentrate on automatic generation of 3D solids that are defined by three orthographic views, that is, the data conversion from 2D to 3D [13] [12]. There are many works reported on this subject. The paper by Weiss and Dori in [4] cites many papers in this field.

2 Two Approaches for 3D Solid Reconstruction

Two approaches are examined in 3D solid regeneration. The first approach is to identify closed loops of line segments in each view and combine them to form multiple intermediate 3D objects called primitives [13]. The target object is then constructed from a set of primitives using appropriate algebraic operations according to a chosen sequence. This constitutes the CSG representation of solid objects. The 3D algebraic operations chosen for this technique are + (addition) and - (subtraction). Each primitive has an attribute that indicates the operation for that primitive. The determination of the attribute is based on closed loop attributes. These are in turn determined by a set of heuristic rules based on relative locations and inclusion relationships of closed loops. Attributes of three closed loops are combined to determine the attribute of the corresponding primitive. The order of operation is determined by a tree structure generated from the original drawing. This technique was successfully applied to a set of drawings to reconstruct 3D objects. With a few sample drawings, however, the technique ran into problems of miss-assigning attributes to primitives and of generating either spurious primitives or not enough primitives.

A major cause of these problems of the first approach is the lack of feedback. The processing is unidirectional and there is no mechanism to check the reconstruction result. The second approach reflects those observations and utilizes 3D operations on 3D objects. This approach is termed "top-down" and the old approach "bottom-up" because of the apparent difference in direction of the reconstruction process. In the first approach, the processing starts from identifying closed loops in the original drawing and works its way up to a 3D object; thus the word "bottom-up". On the other hand, the second approach starts from constructing a 3D object and operates on it until its three views match with the given drawing; thus the notation "top-down". This approach has an explicit mechanism of feeding back the results of reconstruction and correcting discrepancies between what is reconstructed and what is drawn on the given three views. Examples of works adopting this approach are [1][2]. The first phase of our second approach is reported here.

3 Top-Down Approach

First, the type of drawing considered in this study is clarified. The technique discussed is applicable to drawings that are composed of three views with no abbreviations or errors. Boundaries of objects must consist of planes and whole or parts of cylinders. This technique is not currently applicable to cones or spheres. This method will generate a single solid object even if the original drawing can have multiple interpretations.

The basic steps of our top-down approach are the following:

1. Construct three 3D initial objects by sweeping the outermost closed loops of three views along the directions orthogonal to the view planes.
2. Take the product (intersection) of three initial objects to form the starting solid.
3. Draw a set of three regenerated views from the obtained intermediate 3D solid.
4. Compare the regenerated views with the original ones to find "surplus lines" that are not in the original views, "missing lines" that are in the original views but not in the regenerated ones, and "wrong-attribute lines" that have incorrect attributes (e.g., hidden versus solid lines).
5. If there are no surplus lines, missing lines, or wrong-attribute lines then stop. Otherwise go to (6).

6. Determine the next 3D operation for the intermediate 3D solid from the information of how surplus and missing lines are located in three regenerated views.

7. Perform the determined 3D operation on the intermediate 3D solid and loop back to (3).

The flow chart of the reconstruction process is given in Fig. 1.

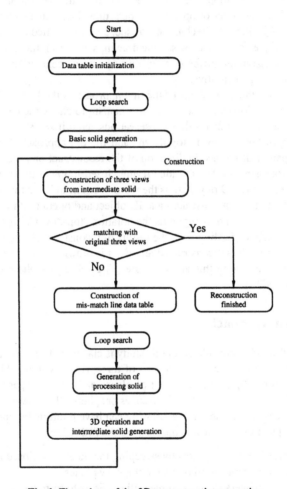

Fig. 1. Flow chart of the 3D reconstruction operation.

Steps (1) and (2) together generate a 3D solid from which the target object is in a way "carved out". Accordingly, the basic 3D operation performed on intermediate 3D solids in step (7) is subtraction. In the prototype system, series of 3D operations are visualized by sequentially showing intermediate 3D solids.

4 Determination of 3D Operations

The major part of this approach is Step (6) above. There are three major processing steps that are recursively used. They are responsible for; (i) searching for closed loops, (ii) generating three views from 3D solids, and (iii) breaking line segments. (i) is the basic operation necessary for forming 3D solids, called processing solids, that are used for carving operation. (ii) is the basic step of feeding the information on the reconstructed 3D solids back to the original three view drawings. (iii) is necessary after (ii) is performed for a number of reasons. There can be multiple lines overlapping each other on the reconstructed three views. Those lines must be broken and regrouped into a set of non-overlapping line segments. An example of this case is depicted in Fig. 2.

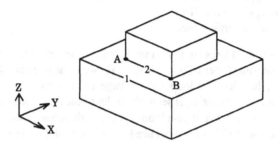

Fig. 2. An example of overlapping line segments that need to be broken and regrouped. Line 2 overlaps with line 1 in the Y-directional view and needs to be separated from line 1.

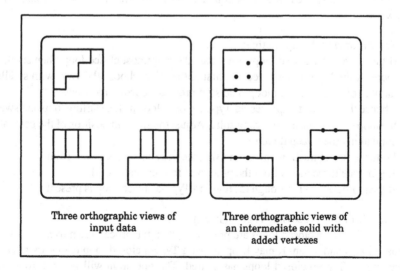

Three orthographic views of input data

Three orthographic views of an intermediate solid with added vertexes

Fig. 3. Adding vertexes to regenerated three views.

Another reason is that a single line segment in the reconstructed views may correspond to a set of two or more line segments in the original three views or vice versa. The heart of the operation in (iii) is to add missing vertices as shown in Fig. 3. Details of operations involved in these steps are discussed in the following subsections.

4.1 Closed loop types

First, closed loops in three views that contain surplus or missing lines are searched and are categorized into five basic types. They are:

a. loops consisting of both surplus and missing lines (and possibly original lines),
b. loops consisting of missing and original (outer) lines,
c. loops consisting of missing and original (internal) lines,
d. loops consisting of missing lines only,
e. loops consisting of surplus lines only.

Here, the original (outer) lines are those that constitute the outer loop of the views. All others are called original (internal) lines. If there is more than one outer loop, such as in the case of a hole, then the outer loop with the largest size is called the outermost loop.

The priority for the processing is set as the order of the type. This ordering is based on the heuristic reasoning that Type-a loops appear in three views when there are portions of the reconstructed 3D solids that need to be removed. Similarly, the loops with higher priority are more likely to represent excess parts of the regenerated 3D solids.

4.2 Processing solids and 3D operation

The approach of generating processing solids and recursive operations using the generated solids is described in this section. The basic algorithm can be summarized as follows:

1. Search for closed loops in three views.
2. If there are outer closed loops other than the outermost closed loop, then set those loops as the loops to be processed first. Otherwise, choose the view with smallest number of Type-a closed loops and start processing from that view.
3. Choose the smallest unprocessed Type-a closed loop as the starting loop and sweep the loop to generate a processing solid. Adjust the size and position of the generated solid along the sweep direction.
4. Perform the subtraction operation using the generated processing solid.
5. Regenerate three views from the processed intermediate solid.
6. Repeat processes (1) through (5) until no Type-a closed loop is present.

An example of this processing is given in Fig. 4.

When all Type-a closed loops are processed, then the processing moves onto Type-b closed loops and so on. It may happen that a Type-a closed loop reappears after the processing of Type-b closed loops has started. The operation will go back to Type-a closed loops in such cases.

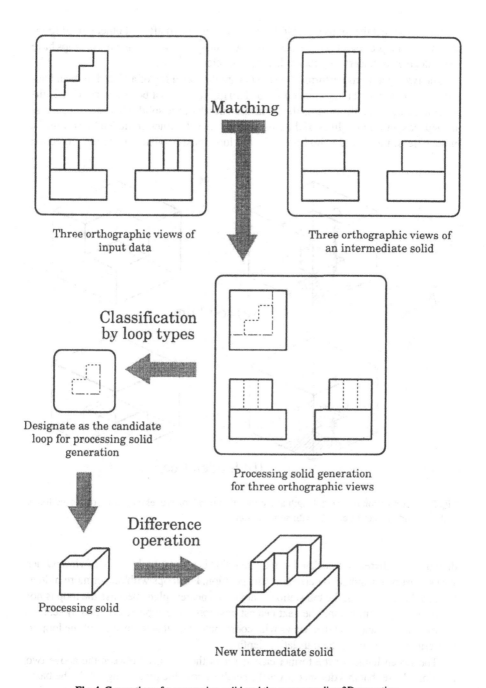

Three orthographic views of
input data

Three orthographic views of
an intermediate solid

Matching

Classification
by loop types

Designate as the candidate
loop for processing solid
generation

Processing solid generation
for three orthographic views

Difference
operation

Processing solid

New intermediate solid

Fig. 4. Generation of a processing solid and the corresponding 3D operation.

It is noted that the loops with hidden lines are processed after all other closed loops of that type are processed. This is based on the heuristic reasoning that those loops have larger degrees of uncertainty than other types of closed loops.

The two operations performed in step (3) are the sweeping of a closed loop and adjusting the position of the generated solid. Two questions must be answered. One is the sweeping length and the other is the translation depth of the solid. This part of processing depends on whether lines of the closed loops are solid lines or hidden lines (Fig. 5). In the case of those with solid lines only, the loop must be visible from the projection

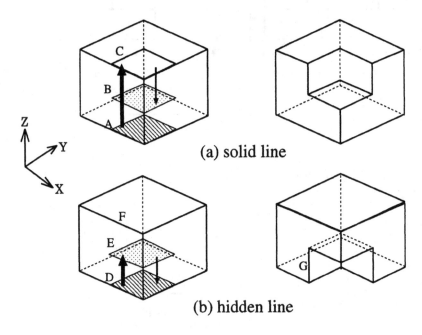

(a) solid line

(b) hidden line

Fig. 5. Determination of sweep length and translation depth of processing solid for closed loops with (a) solid lines only, and (b) with some hidden lines.

direction and therefore translated to the plane that is nearest to the point of view. Other views must be consulted to determine this position. For loops with one or more hidden line, on the other hand, the translation is not to the nearest plane because the loop is not visible from that direction. The next nearest position that can be determined by searching for line segments in other views with coordinates that match the ones of the loop in concern is chosen to be the translation depth.

The sweep length of the former case is set as the distance between the above two positions. Note that this distance gives the smallest possible processing solid. The basic reason for this is the subtraction operation. The chance of avoiding addition operation due to excessive carving can be minimized by this choice. Similarly, the size of the processing solid for closed loops with hidden lines is chosen to be the smallest. Again, this distance can be found by checking other views with line segments with matching coordinates.

5 Application to Examples

The proposed method was coded on a work station (Silicon Graphics Indigo2 Extreme) using DESIGNBASE [3] as the 3D solid modeler software. The method was tested with a set of 45 sample drawings. 40 of them were used in our bottom up approach and the rests were new ones. An example is given in Fig. 6. Out of those 45 samples, the method

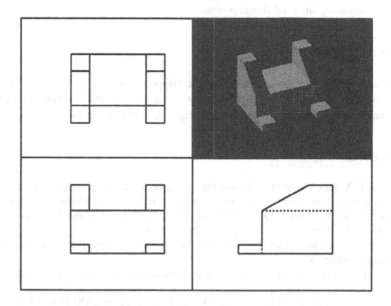

Fig. 6. A sample of three views and reconstruction result. Figures are the top view, front view, side view, and the reconstructed object from top left in the counter-clockwise order.

correctly regenerated 3D solids for 21 samples. There were four other samples where correct 3D solids were regenerated but the reconstructed three views contained lines with wrong attributes.

In all other unsuccessful cases, the proposed method goes into infinite loops of operation. This happens when the same processing solid is generated in two consecutive steps. In the proposed prototype, the subtraction operation is changed to the addition operation when the same processing solid is generated. The infinite looping occurs when this procedure gives rise to conditions that the same closed loop is chosen as the next starting loop. This rule needs to be reconsidered. The history of operations with the corresponding processing solids must be kept and checked in order to alleviate this condition. This method is currently being pursued.

6 Conclusions

A top-down approach of reconstructing 3D solid objects from three view engineering drawings is discussed. The processing starts with generating a 3D solid that contains the target object and then carving out the target solid by generating processing solids from discrepancies between the three views of regenerated 3D solids and the original ones. The method was successfully tested with several examples to show the validity of the approach. The problem of infinite loop of operations was pointed out and a possible way of eliminating this condition is proposed.

Acknowledgements

This research was partially supported by the Nippon Steel Corporation and the Research Institute of Aoyama Gakuin University. The authors express their gratitude to K. Nishimatsu and Y. Kaneko for their help in preparing this manuscript.

References

1. Agui, T., Y. Mesaki, and M. Nakajima 1989. "A study of Reconstructing 3-D Objects from Orthographic Views," Proc., Ann. Spring Mtng. Soc. Elec. Info. and Comm., D-515, 6-235.
2. Chen Z., and D.B. Perng 1988. "Automatic Reconstruction of 3D Solid Objects from 3D Orthographic Views," Pattern Recognition, 21, 5, 439-449.
3. Chiyokura, H. 1988. Solid Modelling with DESIGNBASE: Theory and Implementation, Addison-Wesley, Mass.
4. Dori, D., and A. Bruchstein 1995. Shape, Structure and Pattern Recognition, World Scientific.
5. Idesawa, M. 1986. "3D Model Reconstruction and Processing for CAE," Proc. 8th Int. Conf. Pattern Recog., 220-225.
6. Kitajima, K., and M. Tasaka 1992. "A Method to Reconstruct a CSG Solid Model from a Set of Orthographic Three Views," Trans. Soc. Elec. Info. and Comm., J75-D-II, 9, 1526-1538.
7. Lysak, D.B. Jr., and R. Kasturi 1991. "Interpretation of Engineering Drawings of Polyhedral and non-polyhedral Objects," ICDAR, 79-87.
8. O'Gorman, L., and R. Kasturi 1995. Document Image Analysis, IEEE Computer Society Press.
9. Ota, J., T. Koezuka, H. Arita, T. Nakamura, and K. Tomiyama 1991. "Automatic Recognition of Mechanical Engineering Drawings," Proc. Int. Conf. Ind. Elec. Control and Instrum., Kobe, Japan, 1264-1269.
10. Ota, J., T. Koezuka, H.Arita, T. Nakamura and K. Tomiyama 1994. "Automatic Conversion of Mechanical Engineering Drawings to CAD Data," J. Japan Soc. Precision Engr., 60, 4, 524-529.
11. Senda, T. 1990. "Reconstruction of Solid from a Set of the Orthographical Three Views -Applications to Many Polyhedral Solids-," J. Info. Proc. Soc. Japan, 31, 9, 1312-1320.
12. Shibamiya, T., K. Tomiyama, K. Nakaniwa, T. Yokota, Y. Koezuka and N. Moriuchi 1993. "Automatic Reconstruction of 3D Solid Model from Three Orthographic Views," Info. Proc. Soc. Japan, CG-63, 91-98.
13. Tomiyama, K., and T. Nakamura 1992. "Auxiliary Lines in Three View Drawings for 3D Shape Reconstruction Using CSG Method," Proc. IEEE Int. Conf. Systems Engr., 250-256.

14. Yokoyama, M, and K. Satoh 1993. "Computer Processing of Machine Drawings," Graphics and CAD, 63, 10, 67-73.
15. Yoshiura, H., K. Fujimura and T.L. Kunii 1984. "Top-Down Construction of 3-D Mechanical Object Shapes from Engineering Drawings," IEEE Computer, 17, 12, 32-40.

A Benchmark: Performance Evaluation of Dashed-Line Detection Algorithms

Bin Kong[1], Ihsin T. Phillips[2], Robert M. Haralick[1],
Arathi Prasad[3], and Rangachar Kasturi[3]

[1] Department of Electrical Engineering
University of Washington
Seattle, Washington 98195, USA

[2] Department of Computer Science
Seattle University
Seattle, Washington 98122, USA

[3] Department of Computer Science and Engineering
Pennsylvania State University
University Park, Pennsylvania 16802, USA

Abstract: This paper describes a protocol for systematically evaluating the performance of dashed-line detection algorithms. It includes a test image generator which creates random line patterns subject to prespecified constraints. The generator also outputs ground truth data for each line in the image. The output of the dashed line detection algorithm is then compared to these ground truths and evaluated using a set of criteria.

1 Introduction

Systems which convert existing paper-based engineering diagrams into electronic format are in demand and a few have been developed. However, the performance of these systems is either unknown, or only reported in a limited way by the system developers. A formal evaluation for these systems, or their subsystems, would contribute to the advancement of the field. Since many drawings include some form of dashed lines, we present a dashed-line generation protocol and a performance evaluation protocol for evaluating dashed-line detection algorithms [1]. The benchmark is designed to be used by the recognition system researchers and developers for testing and enhancing their dashed-line recognition algorithms. Figure 1 shows an object-process diagram of the benchmark.

Users of our benchmark first need to use the dashed-line generator to generate a set of test images of various complexities. The generator also produces the ground truth for each image it generates. The algorithm being evaluated operates on these images to perform detection and produces output in a pre-defined format. Next, the performance evaluator takes the detected lines produced by the algorithm and the corresponding ground truth lines produced by the generator and performs evaluation based on a set of criteria. The results of the evaluation are displayed in tabular form.

Note that the contents of these evaluation tables are computed facts. No scores are assigned to algorithms, since assignment of scores to algorithms requires a definition of weights associated with each of the different types of errors. The values of these weights depend on the particular application. In this paper we do not discuss these weights.

This paper is organized as follows: Section 2 contains the procedures for the generation of dashed-line test images. The performance evaluation protocol and its output, performance evaluation tables, are described in Section 3 followed by conclusions in Section 4. Several appendices describe in detail the conventions used and the specifications of various data sets and file formats.

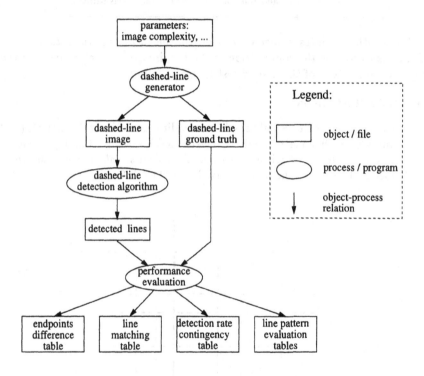

Fig. 1. The object-process diagram of the benchmark. The user supplies the dashed-line generator image complexity parameter. The generator produces a dashed-line image and the corresponding dashed-line ground truth. Detection algorithms perform detection of dashed lines in the image and produces detected lines. The performance evaluator takes the detected lines produced by the algorithm and the ground truth produced by the generator, and performs evaluation. The results of the evaluation are displayed in tabular form.

2 Dashed-line Test Image Generation

Dashed-line detection algorithms are tested using a set of images generated by the Dashed-line Test Image Generator described in this section. Standard graphics generation procedures are used by the generator to create a variety of dashed-lines in various positions and orientations in the image. The parameters controlling the generation procedures are randomly varied to create a rich variety of test images. There are four basic types of line patterns generated. These are shown below:

Solid line   Single-dashed line   Double-dashed line   Dash-dot line

---

Lines within each test image vary in length, thickness and orientation. These variations as well as the composition of the test images are determined by the degree of complexity of the desired test image as follows:

- **Simple Test Image**

  A *simple* image includes only single-dashed lines in horizontal, vertical, (and possibly ±45° diagonal) directions. The dashed-line segment and gap lengths may vary up to ±10% within each line. A typical *simple* image is shown in Figure 2. The size of the image is fixed at 1000 × 1000 pixels.

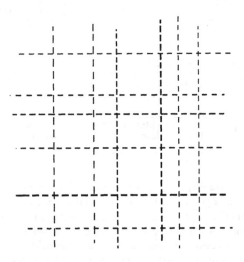

Fig. 2. An image with only horizontal, vertical, and (possibly) diagonal dashed lines representing the "simple" class of images.

- **Medium Test Image**

  In addition to the line patterns included in *simple* images, *medium* complexity images include lines with arbitrary orientation. All three types of dashed lines are included as well as polygons with and without hatching. A typical medium complexity image is shown in Figure 3. Maximum image size is 4000 × 4000 pixels. The segment and gap length parameters may vary by as much as ±40% of the nominal values.

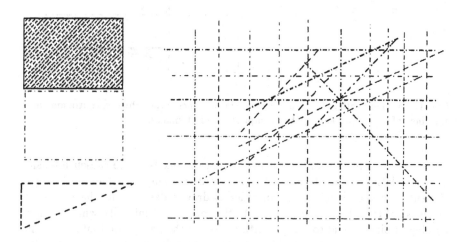

Fig. 3. An image with straight dashed lines and other objects representing the "medium" class of images.

- **Complex Test Image**

  These images include everything in a *medium* complexity image. In addition, curved lines and circles are added and annotations are superimposed on the graphics to further increase the complexity of the image. The length parameters could vary up to ±100% of their nominal values. A typical complex test image is shown in Figure 4.

For a complete specification of the three classes of test images see Appendix 1. All test images are saved by the generator in 8-bit binary TIFF format. More details on the dashed-line primitives generation are given in the following sections.

2.1 Generation of Straight Lines

Line orientation, line length, and the start coordinates of the line are chosen randomly based on the seed input by the user. Once the start and end coordinates of the line are found, the task is now one of determining the points along this

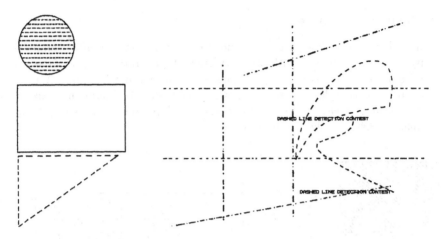

Fig. 4. An image with straight and curved dashed lines, other objects, continuous lines and lines of text representing the "complex" class of images.

line in terms of pixel coordinates and setting the pixels to the chosen intensity. The calculation of points along the line is done using the midpoint form of Bresenham's line algorithm [2]. In order to draw a dashed line, the number of pixels equal to a dash length are set to the foreground intensity while the next number of pixels equal to a gap length are set to the background intensity. This alternate turning "on" and "off" of the pixel intensity is repeated for an entire length of a complete dashed line.

After this stage, all the points along the dashed line are obtained. The next stage is thickening the line. The line is thickened by enclosing it in a rectangle and filling the area. The breadth of the rectangle is set to the line thickness. The rectangle is filled with the foreground intensity by a recursive seed fill algorithm [2]. The seed is taken to be any point central to the area to be filled.

2.2 Generation of Curves

A curved line is drawn as a Bezier curve [2]. The dashed nature of the curve is obtained by toggling the pixel intensities. Four initial control points are chosen randomly and the curve is approximated through these points. The iterative algorithm has the index varying between 0 and 1 in steps of a small, fixed fraction where each succeeding point is calculated as a weighted sum of Bernstein polynomials,

$$x = c_{0x}(1.0 - t)^3 + 3c_{1x}t(1.0 - t)^2 + 3c_{2x}t^2(1.0 - t) + c_{3x}t^3 \qquad (1)$$

$$y = c_{0y}(1.0 - t)^3 + 3c_{1y}t(1.0 - t)^2 + 3c_{2y}t^2(1.0 - t) + c_{3y}t^3 \qquad (2)$$

where x and y are the points on the curve; t is the index of the iteration; and $c_0, c_1, c_2,$ and c_3 are the four control points.

2.3 Generation of Circles

The points on a circle are determined using Bresenham's circle algorithm [2]. Each iteration produces eight equidistant points along the circumference until the desired resolution is achieved. A second circle with the same center but a decreased radius is drawn and the area between the two circles is filled to represent the finite thickness.

3 Performance Evaluation Protocol

When the dashed-line test image generator produces a test image, it also produces, in a predefined format, the corresponding *ground truth* for each line it generates. We require that the output from a dashed-line detection algorithm conform to this predefined format. Detailed specifications of this format are given in Appendix 2. The performance evaluation steps are given below:

1. We assign unique IDs (from $1, 2,...$) to the ground truth lines and the detected lines in the order in which they appear in the files.
2. We compute the *matches* between the ground truth lines and the detected lines regardless of their line types. If there is an offset (a horizontal shifting and/or a vertical shifting) from the detected lines to the ground truth lines, the detected lines are shifted accordingly and the matches are recomputed. The results of matching are output as several tables by the evaluation algorithm. The formats for these tables are described in sections 3.2 and 3.3.
3. For each detected line that matches a ground truth line, we check whether its given line type is correct. The set of detected lines with correct line types is collected for *line pattern* evaluation. We call this set of detected lines the *correctly-matched lines*. For each of the correctly-matched line pairs, we evaluate its line pattern and an evaluation table for each line category is computed (described in section 3.4). No further evaluation is given to any detected line having an incorrect line type.

3.1 Matching Criteria

Our performance evaluation is based on matching the output of the algorithm against the corresponding ground truth of the test image. We consider that a detected line d matches a ground truth line g if:

1. The *angle* between the detected line d and the ground truth line g is less than 3 degrees,
$$angle(d, g) \leq 3$$

2. The *distance* between lines d and g is less than 5 pixels,
$$llDist(d, g) \leq 5$$

3. The *Relative overlap* function of d and g with respect to the orientation of g is at least 0.8,

$$Reloverlap(d, g, \alpha_g) > 0.8, \text{where } \alpha_g = orient(g).$$

For a precise definition of these functions see Appendix 3. The thresholds were determined heuristically based on the properties of the test images such as dashed line thickness, line separation, minimum line length, etc.

A problem arises if a detected line matches multiple ground truth lines, and/or if there are other detected lines which also match the same ground truth line. In the first case, it is not correct to consider that the same detected line matches all ground truth lines. In this case, we only match the detected line to that ground truth line with which it has maximum *relative overlap*; other ground truth lines are considered as misdetections. In the second case, it is not correct to consider multiple detections of the same ground truth target as independent correct detections. Hence, only the detected line having the largest *relative overlap* with the ground truth line is considered matched with the ground truth; other detected-lines are considered as false alarms.

3.2 Offset Evaluation

During offset estimation, only matched detected-lines are considered. For each matched detected-line, the rows and the columns differences between its two endpoints and the two endpoints of its corresponding ground truth are computed. Note that there are four values here, i.e., a row and a column difference for each pair of matched endpoints. There will be two sets of the row differences and two sets of column differences after all matched lines are considered. The means and the variances for each of these four sets are estimated as follows:

1. Compute the initial mean and the initial variance for the set.
2. If a value within the set is greater than twice the estimated standard deviation of the set, the value is eliminated from the set. This step helps to group only those detected lines which are uniformly offset from their corresponding ground truth lines by avoiding the influence of outliers.
3. Compute the new mean and the new variance for the set.
4. Repeat steps 2 and 3 until the values become stable.

Finally, the offset estimation is computed using the four final means and the four final variances. Recall that there are two row means and two column means in the four sets. If the smaller of the two row variances is less than a predefined threshold value (we use 4.0 here), then the corresponding mean values are considered as the offset in row values. Otherwise, we consider that there is no offset. The column offset is computed in a similar fashion.

All this information is maintained in the *endpoints difference table*. The entries of the *endpoints difference table* give the differences, in term of columns and rows, between the endpoints of the *detected lines* and the endpoints of the matched *ground truth lines*. The estimated mean and the estimated variance of the offsets are displayed in the last two rows of the *endpoints difference table*. If for some reason the *detected lines* are uniformly shifted either horizontally and/or vertically, the shift will be shown in the *endpoints difference table*, i.e., all column entries and/or all row entries will be identical. If a global offset is detected, the entire set of the detected lines are shifted, and the matches are recomputed.

3.3 Line Match Evaluation

A *match table* is also created during *line match evaluation*. A *match table* entry is either a 1 (a match) or a blank (no match). A 1 at the entry (i, j) indicates that the ith *detected line* matches the jth *ground truth line*. Within the *match table*, we also include misdetections and false-alarms. The entries for the ground truth misdetections are given in the last row of the *match table*. The entries for the falsely detected lines are given in the last column of the *match table*. A 1 on jth entry of the misdetection row indicates that the jth *ground truth line* was not detected by the algorithm. A 1 on ith entry of the false-alarm column indicates that the algorithm produced a line which is not a part of the ground truth.

Figure 5 contains templates of the two *contingency tables*. The top table contains the total number of *ground truth lines* which have been labeled as each of the line types: Solid line, Dashed line, Double-dashed line, and Dash-dotted line. The bottom table contains the correct-detection rate, the mis-label rate, the misdetection rate, and the false-alarm rate for each of the line types. These are calculated using the following equations:

$$P_{\text{correct}} = \frac{(A_1 + B_2 + C_3 + D_4)}{N_g} \tag{3}$$

$$P_{\text{mis-lab}} = \frac{\sum_i (a_i + b_i + c_i + d_i)}{N_g} \tag{4}$$

$$P_{\text{mis-detect}} = \frac{\sum_i M_i}{N_g} \tag{5}$$

$$P_{\text{false}} = \frac{\sum_i F_i}{N_d} \tag{6}$$

where the various quantities are shown in the top table of Figure 5 and the values N_g and N_d are the total number of ground truth lines and total number of detected lines, respectively.

3.4 Line Pattern Evaluation

Figure 6 contains a template of the *Single-dashed line pattern evaluation table*. The table displays the mean and variance of the dash length and the mean gap length for each matched pair of *ground truth* and *detected* lines. Chi-squared values ($\sum \frac{(D-G)^2}{G}$) are also computed. Similar tables are also created for other types of dashed lines.

Detected Lines

		Solid	Double-dash	Single-dash	Dash-dot	Mis-detect
	Solid	A_1	a_2	a_3	a_4	M_1
	Double-dash	b_1	B_2	b_3	b_4	M_2
	Single-dash	c_1	c_2	C_3	c_4	M_3
	Dash-dot	d_1	d_2	d_3	D_4	M_4
	False alarm	F_1	F_2	F_3	F_4	

(Left label: **Ground Truth Lines**)

$A_1, a_i = $ number of ground truth Solid lines detected as Solid, Double-dash, etc.
$B_2, b_i = $ number of ground truth Single-dash lines detected as Solid, Double-dash, etc.
$C_3, c_i = $ number of ground truth Double-dash lines detected as Solid, Double-dash, etc.
$D_4, d_i = $ number of ground truth Dash-dot lines detected as Solid, Double-dash, etc.
$M_i = $ number of ground truth lines of Solid, Double-dash, etc. that were misdetected.
$F_i = $ number of Solid lines, Double-dash lines, etc. that were falsely detected.

Detected Lines

		Line-type i	Not Line-type i	Not detected
	Line-type i	$P_{correct}$	$P_{mis\text{-}lab}$	$P_{mis\text{-}det}$
	Not ground truth	P_{false}		

(Left label: **Ground Truth**)

Fig. 5. The top table contains the number of *ground truth lines* which are labeled as Solid lines, Dashed lines, Double-dashed lines, and Dash-dotted lines. The bottom table contains the correct-detection rate, the mis-label rate, the mis-detection rate, and the false-alarm rate.

4 Conclusions

A formal protocol for systematically evaluating the performance of dashed-line detection algorithms was described in this paper. A test image generator which generates a rich variety of test images for evaluating dashed-line detection algorithms was presented. The evaluation procedures as well as the test images, including source code, were made available for ftp well in advance of the contest. The participants were required to run their software for dashed-line detection on a new set of test images generated just before the contest. One of the limitations of this contest is that the testing was limited to synthetically generated images. This was done so that random images with accurate ground truth data could be

Single-dashed Line Pattern Evaluation Table

Index		Dashes				Gaps	
		Mean		Variance		Mean	
G	D	G	D	G	D	G	D
g_{21} g_{22} : : : g_{2b}	d_{21} d_{22} : : : d_{2b}						
Chi Square							

Fig. 6. The table displays the mean and the variance of the lengths of dashes and the mean of the lengths of the gaps for each matched pairs of the *ground truth line* and the *detected line*.

generated just before testing to ensure that the images used in testing would not be known to anyone. Testing on scanned paper documents would have required manual ground truthing prior to the contest. Clearly, for a more rigorous testing in a non-contest environment it is essential to test on scanned document pages from a variety of sources. Another limitation of the evaluation protocol is that it does not evaluate matching of dashed lines formed by circles and other curves. Such matches were subjectively evaluated.

Acknowledgements

We would like to acknowledge the help of David Kosiba for his help in organizing this contest. We would like to thank the graphics recognition community for its keen interest and its input on the format and organization of this contest. In particular, we are indebted to Karl Tombre for his enthusiasm, insight and support for this activity. We also thank the reviewers for many helpful suggestions in enhancing the readability of this paper.

Appendix 1: Complete Specifications of Test Image Classes

A1.1 Simple Test Image

a. The image size is 1000 × 1000 pixels.
b. Only *single-dashed lines* are used.
c. Only horizontal, vertical, and diagonal (±45%) dashed lines.
d. A minimum distance of 50 pixels between lines.
e. Minimum number of dashed lines is 10 while the maximum number is 20.
f. Segment length varies from 10 - 30 pixels.
g. Gap length varies from 1 - 10 pixels.
h. Segment length to gap length ratio varies from 0.8 - 2.0.
i. Thickness of a dashed line varies from 3 - 30 pixels.
j. The above parameters (f, g) have variations of the order of ±10%.
k. Minimum length of a dashed line is 50 pixels.

A1.2 Medium Test Image

a. The image size varies from 1000 × 1000 to 4000 × 4000.
b. All three kinds of dashed lines (*single-dashed, double-dashed, and dash-dot*) are present.
c. Lines can be in any random orientation, but only 4 orientations possible in an image (Note that all lines are straight lines).
d. Angles in an image have a minimum difference of 20 degrees between them.
e. Minimum number of dashed lines is 20 and the maximum number is 40.
f. The segment and gap-length parameters have variations of up to ±40%.
g. All other specifications are same as those for Simple images.

A1.3 Complex Test Image

a. The image size varies from 4000 × 4000 to 8000 × 8000.
b. In addition to the three kinds of dashed lines, solid lines and text may also be present.
c. Both straight and curved dashed lines are possible.
d. Minimum distance between lines equal to the thickness of lines.
e. Minimum number of dashed lines is 30 while the maximum number is 100.
f. Segment length to gap length ratio varies from 0.8 - 4.0.
g. The segment-length parameters could vary up to ±100% and missing segments are possible.
h. All other specifications are same as those for Medium images.

Appendix 2:
Output Specification for Dashed-line Detection Algorithms

The output of the dashed line detection algorithms should conform to the following specifications:

1. One text-line for each detected line.
2. For each detected line, the output text-line is of the format:

$$n \ c_1 \ r_1 \ c_2 \ r_2[additional\text{-}parameters]$$

   where n is 1 for solid line, 2 for single-dashed line, 3 for double-dashed line, and 4 for dash-dot line and c_1, r_1 and c_2, r_2 are the column and the row positions of the first and the second end points of the detected line. We require that $c_1 < c_2$. If $c_1 = c_2$, then $r_1 < r_2$.
3. For a Single-dashed line, the output text-line includes the following additional parameters: mean-dash-length, dash-variance and mean-gap-length. Note that, the *gap-length variance* will be identical to that of *dash-variance*, therefore, the *gap-length variance* is not required.
4. For a Double-dashed line, the output text-line includes the following additional parameters: *mean-dash1-length, dash1-variance, mean-dash2-length, dash2-variance, and mean-gap-length.* Length of dash-1 should be longer than that of dash-2.
5. For a Dash-dot line, the output text-line includes the following additional parameters: *mean-dash-length, dash-variance, mean-dot-width, width-variance, and mean-gap-length* where *mean-dot-width* is the average diameter of the dots.

Appendix 3: Conventions and Definitions

In this appendix, we give the conventions and definitions that are used in this work.

A3.1 Image Coordinate System

An image is given by columns and rows of pixels. In an 8-bit binary image, a foreground pixel has the value 255 and a background pixel has the value 0. We use the *column-row*[1] coordinate system, (c-coordinate, r-coordinate), to represent a pixel's position within an image. The origin of this system, $(0, 0)$, is at the top-left corner pixel of the image.

---

[1] We use the column-row notation to avoid the confusion caused by the orientation differences in the (x, y) Cartesian system notation and the $[i, j]$ image array notation.

A3.2 Definitions

- **Line orientation**

  The orientation of a line segment $l = (c_1, r_1, c_2, r_2)$, denoted as $orient(l)$, is the angle between l and the c-axis (see Figure 7). The range of $orient(l)$ is $(-90°, 90°]$. The function $orient(l)$ is

$$orient(l) = \begin{cases} \arctan\left(\dfrac{r_2 - r_1}{c_2 - c_1}\right), & \text{if } c_1 \neq c_2 \\ 90°, & \text{if } c_1 = c_2 \end{cases}$$

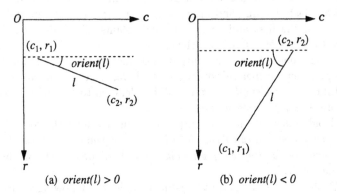

(a) $orient(l) > 0$ (b) $orient(l) < 0$

Fig. 7. The orientation of a line segment.

- **Angle between two line segments**

  The angle between two given line segments l_1 and l_2, is defined as the included angle between l_1 and l_2, and denoted as $angle(l_1, l_2)$ (see Figure 8). The range of $angle(l_1, l_2)$ is $[0°, 90°]$. The function $angle(l_1, l_2)$ is

$$angle(l_1, l_2) = \begin{cases} |orient(l_1) - orient(l_2)|, & \text{if } |orient(l_1) - orient(l_2)| \leq 90° \\ 180° - |orient(l_1) - orient(l_2)|, & \text{otherwise} \end{cases}$$

- **Point-line distance**

  The point-line distance between a point $p = (c, r)$ and a line segment $l = (c_1, r_1, c_2, r_2)$, is the orthogonal distance from p to l and denoted as $plDist(p, l)$ (see Figure 9). The point-line distance function is

$$plDist(p, l) = \frac{|(c_2 - c_1)r - (r_2 - r_1)c - (r_1 c_2 - r_2 c_1)|}{\sqrt{(c_2 - c_1)^2 + (r_2 - r_1)^2}}.$$

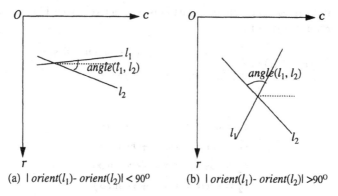

(a) $| orient(l_1) - orient(l_2)| < 90°$ (b) $| orient(l_1) - orient(l_2)| > 90°$

Fig. 8. The angle between two line segments.

- **Line-line distance**
  Let $l_1 = (c_{11}, r_{11}, c_{12}, r_{12})$ and $l_2 = (c_{21}, r_{21}, c_{22}, r_{22})$ be two line segments. The midpoints of l_1 and l_2 are m_1 and m_2, respectively. The line-line distance of l_1 and l_2, $llDist(l_1, l_2)$, is defined as the average of the point-line distance of m_1 to l_2 and that of m_2 to l_1 (see Figure 9). The line-line distance function is

  $$llDist(l_1, l_2) = \frac{1}{2}\left(plDist(m_1, l_2) + plDist(m_2, l_1)\right).$$

  where

  $$m_1 = \left(\frac{c_{11} + c_{12}}{2}, \frac{r_{11} + r_{12}}{2}\right) \text{ and } m_2 = \left(\frac{c_{21} + c_{22}}{2}, \frac{r_{21} + r_{22}}{2}\right).$$

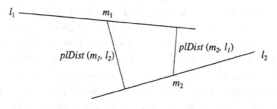

$$llDist(l_1, l_2) = (plDist(m_1, l_2) + plDist(m_2, l_1)) / 2$$

Fig. 9. The line-line distance between two line segments.

- **α-projection of a line segment**
  The α-projection of a line $l = (c_1, r_1, c_2, r_2)$ is the projection of l onto the given orientation $\alpha \in (-90°, 90°]$. The α-projection of l, $proj(l, \alpha)$, is also a line segment. Its two endpoints (c_1', r_1') and (c_2', r_2') are the projections of

(c_1, r_1) and (c_2, r_2) onto the orientation α, respectively (see Figure 10). The α- projection is given by

$$proj(l, \alpha) = \begin{cases} (c_1', r_1', c_2', r_2'), & \text{if } |\alpha - orient(l)| \leq 90^\circ \\ (c_2', r_2', c_1', r_1'), & \text{otherwise} \end{cases}$$

where

$$c_1' = \cos\alpha(c_1\cos\alpha + r_1\sin\alpha)$$
$$r_1' = \sin\alpha(c_1\cos\alpha + r_1\sin\alpha)$$
$$c_2' = \cos\alpha(c_2\cos\alpha + r_2\sin\alpha)$$
$$r_2' = \sin\alpha(c_2\cos\alpha + r_2\sin\alpha).$$

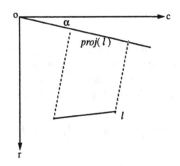

Fig. 10. The α-projection of a line segment.

- **Overlap between two line segments – a relationship function**
  The α-overlap between two line segment l_1 and l_2, $overlap(l_1, l_2, \alpha)$, is a relationship function of l_1 and l_2 with respect to a given orientation α. Suppose T_a and T_d are two given thresholds which are determined by the user based on the application in which the dashed line detection algorithm is used. T_a is a threshold for the angle between two line segments and T_d is a threshold for the line-line distance. If $angle(l_1, l_2)$ is not greater than T_a, and $llDist(l_1, l_2)$ is not greater than T_d, we say that l_1 and l_2 are *sufficiently close* to each other.
  The α-overlap of l_1 and l_2 is defined as the length of the common part of their α-projections if l_1 and l_2 are sufficiently close, and is defined as 0 otherwise. The function $overlap(l_1, l_2, \alpha)$ is

$$overlap(l_1, l_2, \alpha) = \begin{cases} length(proj(\ l_1, \alpha) \cap proj(l_2, \alpha)), \\ \quad \text{if } angle(l_1, l_2) \leq T_a \text{ and } llDist(l_1, l_2) \leq T_d \\ 0, \quad \text{otherwise} \end{cases}$$

- **Relative overlap**

  The relative overlap of two line segments l_1 and l_2 is defined as the ratio between the overlap function $overlap(l_1, l_2, \alpha)$ and the length of the longer segment:

  $$reloverlap(l_1, l_2, \alpha) = \frac{overlap(l_1, l_2, \alpha)}{\max(length(l_1), length(l_2))}$$

References

1. R.M. Haralick, "Detection Performance Methodology" *Proceedings Image Understanding Workshop*, Palm Springs, California, February 1996, Vol II, pp. 981-983.
2. J.D. Foley and A. Van Dam, "Computer Graphics, Principles and Practice" Addison-Wesley, Reading, Massachusetts, 1991

How to Win a Dashed Line Detection Contest

Dov Dori   Liu Wenyin   Mor Peleg

Faculty of Industrial Engineering and Management
Technion—Israel Institute of Technology
Haifa 32000, Israel

{dori, liuwy, mor}@ie.technion.ac.il

Abstract. Correct recognition of dashed lines is essential for high-level technical drawing understanding. Automatic solution is quite difficult due to the limitations of machine vision algorithm. In order to promote development of better techniques, a dashed line detection contest was held at the Pennsylvania State University during the First International Workshop on Graphics Recognition, August 9-11, 1995. The contest required automatic detection of dashed lines on test drawings at three difficulty levels: simple, medium and complex, which contained dashed and dash-dotted lines in straight and curved shapes, and even interwoven texts. This paper presents dashed line detection techniques which won the first place in the contest. It successfully detected the dashed lines in all drawings. The underlying mechanism is a sequential stepwise recovery of components that meet certain continuity conditions. Results of experiments are presented and discussed.

Keywords. Dashed Line, Graphics Recognition, Technical Drawings

1. Introduction

Lines are the most basic graphical primitives in line drawings in general and in technical drawings in particular. Humans find it easy to recognize discontinuous lines because of their well developed visual perception abilities to 'fill the gaps' between consecutive segments. Machines do not possess this natural ability and therefore the task of recognizing discontinuous lines by machines turns out to be rather non-trivial, as we show in this paper. In order to promote development of better techniques, a dashed line detection contest was held at the Pennsylvania State University during the First International Workshop on Graphics Recognition, August 9-11, 1995. The contest required automatic detection of dashed lines on test drawings at three difficulty levels: simple, medium and complex. The drawings are generated randomly and contain dashed and dash-dotted lines in straight and curved shapes, and even interwoven texts. This paper describes our algorithm for detecting such dashed lines.

Several groups have already discussed the dashed line detection problem. Almost all groups use vector-based algorithm to solve this problem, except for Boatto et al. [1], who used a semi-vector-based method to find dash segments having special graph structures. Vaxiviere and Tombre's Celesstin system can detect both dashed lines and dash-dotted lines according to the French Standard NF E 04-103 [2]. Kasturi et al. [3] have done work on detecting dashed lines in drawings and maps. They attempt to recognize dashed lines that are not necessarily straight, as is the case in maps. Joseph and Pridmore [4] have dealt with finding dashed lines in engineering drawings by looking for chains of short lines within their ANON system.

Our dashed line detection algorithm, which is also vector-based, is a specification of the dashed line detection module developed within the Machine Drawing Understanding System—MDUS, which is designed for 2D and 3D understanding of objects described in the engineering drawings. This paper presents the dashed line detection techniques we used in the software for the contest. It successfully detected out the "discontinuous lines" (a generic name we used to avoid confusion with all styles of lines that involve dashes, including dashed lines and dash-dotted lines, and even dash-dot-dotted lines), which have both straight and curved geometry at all three levels of drawings. The underlying mechanism in the techniques is a sequential stepwise recovery of components that meet certain continuity conditions. Results of experiments are presented and discussed.

The contest specifications, characteristics of various classes of test images and the criteria for evaluation are described in a companion paper in this volume [5]. In Section 2 we describe our terminology and the characteristics of discontinuous lines. In Section 3 we describe the detection algorithms for both straight and curved dashed lines. In Section 4, we present a procedure for detecting dashed blocks. Finally, the experimental results are presented in Section 5.

2. The Components of Discontinuous Lines

The terms defined below are used to describe the components and attributes of discontinuous lines in this section and are used throughout the work.

(1) Dash—a bar or an arc, which is part of a discontinuous line.

(2) Dot—a dash whose length is no more than n times its width, where n is a small number (2 or 3).

(3) *Segment*—a dash or a dot. Segments are denoted $S_{[R1]i}$, where i is the number of segments from one end of the discontinuous line. $S_{[R2]i}$ denotes the ith segment of the discontinuous line.

(4) *Gap*—a void straight line segment, whose endpoints coincide with the two close endpoints of the two neighboring segments of the same discontinuous line. A gap between segments $S_{[R3]i-1}$ and $S_{[R4]i}$ (i=2, 3, ..., 5) is denoted $G_{[R5]i}$.

(5) *Center line*—a zero width straight or curved line, which is the medial axis of a segment.

(6) *Deviation Angle*—the angle between a base line and an adjacent line segment, in the direction from the base line to the newly detected line segment. The sign of the angle's value is positive if the deviation is clockwise and negative if it is counterclockwise. AG_i is the angle by which $G_{[R6]i}$ deviates from S_{i-1}, and AS_i is the angle by which $S_{[R7]i}$ deviates from $G_{[R8]i}$, (i=2, 3, ..., 5; see Figure 4).

(7) *Virtual line*—the line formed by joining consecutive segments of a discontinuous line; the trajectory of the perceived line formed by mentally bridging the discontinuities along the discontinuous line.

The most prominent common feature of discontinuous lines is that they are all composed of a number of (at least 2 or 3) segments with a gap between two neighboring ones. The dashes are of approximately equal length. So are the gaps. In CAD systems, both the variance in segment length and the variance in the gap length are naturally much smaller than those in manually prepared drawings. The dots are far shorter than the dashes, and they may appear as either black spots or very short segments (as in the medium and complex test cases). In general, the lengths of the segments and the gaps of discontinuous lines are only limited relatively rather than absolutely. But in the contest requirements, they are limited absolutely by some values and variances. In addition, some other parameters are also confined, such as the ratio between segment and gap lengths, segment thickness and even the minimal length of the discontinuous lines [5]. These limitations definitely facilitate detection, compared with non-restricted cases in real life drawings.

The geometry attributes of discontinuous lines in engineering drawings may have three different values: straight, circular arc, and free curve. In spite of this variability, all discontinuous lines share the following two common features:

• They consist of broken, relatively short line segments.

• All their segments lie along the virtual line.

These two common features are used to detect the class of discontinuous lines regardless of their type. The determination of the type is done on the basis of the observations of the segment length difference:

• Dashed lines have approximately equal length segments.

- Dash-dotted lines have about as many short segments as they have long segments.

- The geometry of the line (straight, circular arc, or free form) is inferred from the shape of the union of the line's segments and the gaps, which, together, form the virtual line.

3. The Discontinuous Lines Detection Algorithm

Our discontinuous line detection starts immediately following vectorization and/or arc segmentation, which extract bars and/or arcs, respectively. We do not need to work with pixel data at all. Any vectorization method (such as OZZ [6,7] and PBT [8]) can be used, as long as the final result is a list of bars and/or arcs, which must be supplied as input to our algorithm. The vectorization method we used is Sparse Pixel Tracking [9]. We did not apply arc segmentation in the program for the contest.

3.1 The Generic Object Recognition Methodology

Our discontinuous line detection algorithm is a specialization of the unified generic object recognition methodology [10], which is briefly summarized below.

The generic recognition algorithm consists of two main phases: (1) hypothesis generation, in which we detect the first key component of the sought object; and (2) hypothesis test, in which we extend the detected component, if possible.

In the first phase we find the first key component of the object to be recognized. Hence, making the correct hypothesis is crucial, and should be properly constrained. If it is over-constrained, only a few objects will be found, while under-constraining it would lead to too many false alarms. If no such key component can be found, it means there are no more objects of the type being sought and the recognition process stops.

The second phase—the hypothesis test—uses the detected key component to extend the detection of the object in the proper directions to the maximum possible extent in order to find all the other components. Here we test the hypothesis put forth in the first step. The components are found gradually, one by one, in areas that extend from the area in which first key component is located. After each component is found, the search area is moved forward in the search direction.

The key component selected in the first phase is tested in the second phase for being part of the hypothesized object. The test succeeds if at least some other meaningful components of the object are detected in their anticipated locations. Success in the test means that an object of the type being sought has been recognized. If the test fails, the key component is rejected as being a part of the anticipated object. Regardless of whether the test is successful or not, the recognition process proceeds to find the next key component to start a new hypothesis test.

Area search is the most commonly used operation in this algorithm. This is a time consuming operation which requires high time efficiency. To realize this methodology, a new data structure, the position index, has been developed and is used to organize the data during the object recognition process [10]. The position index realizes mapping from positions (planar locations) to primitives in the drawing, thereby expediting the search for given objects in a given area.

This generic algorithm can be applied to realize the recognition process of a variety of objects. It serves as a framework within which the specific, syntax-based rules for each object class are applied. In this work, we apply the generic algorithm to discontinuous lines detection.

3.2 Specialization into Discontinuous Line Detection

Being a specialization of the generic object recognition methodology, the discontinuous line detection also consists of two corresponding steps. A bar from the bar list generated by the vectorization, such as segment $S_{[R9]i}$ in Figure 1, with the following attribute values is selected in the first step as the first key component:

1. The bar length has to be within some boundary values, as determined by the specifications of the test data.

2. Both endpoints of the bar have to be free, i.e., no other object may pass through any one of the bar's endpoints.

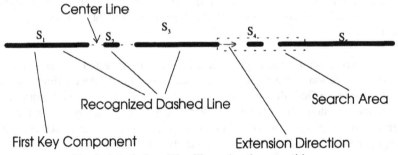

Fig. 1: A dash-dotted line illustrating the recognition.

The extension direction in the second, hypothesis test phase, is determined as going farther away from the new endpoint which is most remote from the first key component while using the same slope as that of the former component. The search area is defined as the rectangle whose width is a small number (2 or 3) times the line width, and whose length is somewhat longer than the maximum allowed gap length in the extension direction. When more than one candidate that satisfies the constraints are found, the nearest constrained one to the most recently detected component is selected. Thus, $S_{[R10]4}$ rather than $S_{[R11]5}$ in Figure 1 is selected as the component that follows $S_{[R12]3}$. The component candidates found in the search area in this step are further constrained by the following three conditions, which are added to the two used in the first step.

3. Each new component should leave a gap from the close edge of the recognized discontinuous line, which is about equal to the cumulative average of the previous gaps' lengths.

4. The new component can be either a long segment, which should have approximately the same width and length as the previous ones, or a short segment, which is only constrained by length, (as its width may be small).

5. The new component should be aligned with respect to the previous components such that it contributes to a meaningful continuity of the corresponding virtual line.

Condition 2, which requires two free endpoints, may be set more loosely at the hypothesis test phase, because the discontinuous line may cross other primitives, (e.g., contour lines), such that the edge of a dash may accidentally fall on some other object.

Constraint 5 is more abstract than the others. The most noticeable difference is between discontinuous straight lines on one hand and discontinuous curved lines (arcs or free curves) on the other hand. Segments of discontinuous lines in real life drawings may also be interwoven with each other, as in Figure 2, which further complicates correct recognition. Next we discuss the detection of straight and curved lines and the specific operations and conditions associated with each geometry.

3.3 Discontinuous Straight Line Detection

For straight discontinuous lines, the segments are usually quite collinear with each other in general, but in practice some short segments may deviate slightly from the center line. The dots may also be shorter than the width of the dash-dotted line. In this case the dot segments may not be collinear with the edges of the longer segments. To overcome this problem, we introduce the constraint that no line segment edge should lie outside the area defined by the width of the line's constituent dashes, extending by w to both sides of the center line of the virtual line, where w is the line width. In the program for the contest, the width of this domain is set with a tolerance of twice the average dash width, as illustrated in Figure 2.

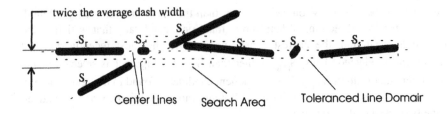

Fig. 2: Illustration of the base line and the line domain of the discontinuous straight line.

Segments $S_{[R13]1}$ through $S_{[R14]5}$ in Figure 4 form a discontinuous straight line, while Segments $S_{[R15]6}$ and $S_{[R16]7}$ may form another discontinuous straight line. Although $S_{[R17]3}$, $S_{[R18]4}$ and $S_{[R19]5}$ are not collinear with the previous components and deviate by a small angle from the center line, they still lie within the line domain. Hence they are still considered components of the discontinuous straight line. While $S_{[R20]6}$ is also found in the search area, one of its endpoints lie outside the line domain of the current discontinuous straight line, so $S_{[R21]6}$ is not recognized as a component of the discontinuous straight line composed of $S_{[R22]1}$- $S_{[R23]5}$.

Since the components of discontinuous straight lines are themselves straight line segments, the constraint exemplified by $S_{[R24]6}$ can be expressed in term of the distance from the segment's endpoints to the center line by the following rule.

> *If the distances of each endpoint of the candidate component from the current center line is less than a tolerance, this candidate is recognized as a component of the discontinuous straight line.*

Since the length of the segment under consideration is irrelevant, the rule is applicable to both types of discontinuous lines: dashed lines, dash-dotted lines.

It may seem that the same constraint can be expressed in different terms, by using, for example, the deviation angle of the segments from the base line. However, if we require that the deviation be less than some threshold, we run the risk of not recognizing dots, because the deviation angle may be very large if the dot segment is very short.

A Final curvature test is needed to determine the discontinuous line geometry, because the cumulative error may be greater than the tolerance but the line may still be erroneously classified as a discontinuous straight line. Such is the case in Figure 3. Although every segment lies inside the line domain defined by the current center line, the final recognized discontinuous straight line is in fact a discontinuous curved line. The final test therefore checks whether each one of the intermediate segments lies within the line width domain of the straight line, whose endpoints are the farthest two points on this discontinuous line. If this condition is not satisfied, the line is curved rather than straight. In the border case of very small curvatures, this curvature test may fail, resulting in the classification of slightly curved discontinuous lines as straight, as we show in the experimental results (see Figure 7).

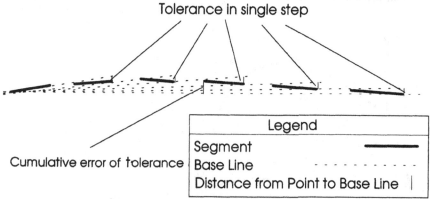

Fig. 3: Illustration of accumulated error of tolerance.

Following the curvature test, classification into one of the two discontinuous line types (i.e., dashed or dash-dotted) is done on the basis of dash length variation. Lines having dashes with approximately small length variance may be classified as dashed lines. Lines having about half long and half short dashes may be classified as dash-dotted. In the program for the contest, we only used the number of short dashes as the criteria. First the dashes are classified as short or long compared to the average length of dashes. Then, the line is classified as dash-dotted if the number of short dashes exceeds both one-fifth of the total dash number in this line and a minimum number of two, otherwise the line is considered as dashed line. This rule reflects the fact that dots may be missed during vectorization and some dots may come from other intersecting dash-dotted lines.

3.4 Curved Dashed Line Detection

The center line of a discontinuous curve is more difficult to estimate than that of a straight one, because consecutive segments are not collinear. This is especially true for cases of free formed curves in the contest image. However, the dash slopes may not be too far apart from each other. For dashed curves, Constraint 5 in Section 3.2 is therefore expressed differently. Here, the deviation angle between the gap of the new component and the previously detected component is used as the main factor in the detection.

Examining Figure 4, we note that every new segment deviates in the same direction (clockwise or counterclockwise) from the nearest previous gap, as the gap deviates from the nearest previous segment.

Based on this observation, Constraint 5 is expressed by the following rule (see Figure 4).

Let T be the maximal allowed deviation angle. If $AG_i * AS_i > 0$ and $AG_i < T$ and $AS_i < T$, then the candidate segment S_i is considered as the next component of the discontinuous curve.

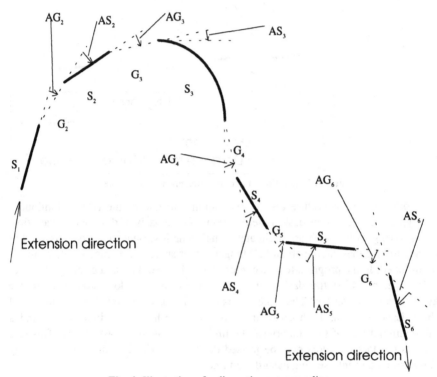

Fig. 4: Illustration of a discontinuous curve line.

In practice, we found that this rule is not applicable to all cases. One of the reasons is that the dashes of the discontinuous curve may be separated by a gap, within which there is one or two dots, as in Figure 5(a), such that the angle AG_i and/or AS_i are too large. As noted, we do not use dots to compute deviation angles, because their slopes are too imprecise. Another reason, shown in Figure 5(b) is the possible misalignment of two consecutive dashes. S_2 extends backwards, causing AG_2 and AS_2 to have opposite signs. In the first stage, only longer segments, rather than shorter segments, are taken into consideration, because the deviation angle of the shorter segment is very imprecise.

To solve the case of dash misalignment, we use a dynamic tolerance of the deviation angle. The tolerance A_t is set relative to AS_i. We use $A_t = AS_i/2$. If $AG_i * AS_i < 0$, then we test whether $(AG_i + A_t) * AS_i > 0$.

The tolerance $AS_i/2$ compensates for the dash misalignment by increasing AG_i in the direction of AS_i, so that AG_i and AS_i may have the same sign. The angle test is applicable only to longer segments of discontinuous curves. If the deviation angle is too large, we search the triangular area between two neighboring dashes, such as in Figure 5(a). If there are some short segments in this area, they can be recognized as dots of the discontinuous curve. The line type is also recognized as dashed curve, dash-dotted curve, or even dash-dot-dotted curve, according to ratio between the number of short and long segments.

After all segments are found, the curve shape of the recognized discontinuous curve line can be inferred by comparing the virtual line to known geometries, such as circular or elliptic arcs. If no such known shape can be matched, the discontinuous curve is regarded as a free form curve. Free form discontinuous curves are found frequently in maps of many kinds.

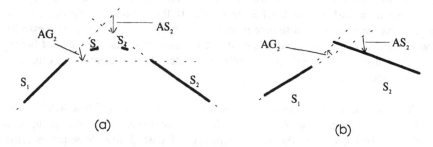

Fig. 5: Problems with dashed curve detection. (a) The presence of dots can cause the deviation angle between dashes to be too large. (b) Extension of S_2 backwards can cause AG_2 and AS_2 to have different directions.

4. The Detection of Dashed Blocks

We used several passes to detect different types of objects. First, straight lines are detected by continuously finding the first key component from the bar list and extending it using the straight line criteria. Curved lines are then detected by finding the first key component from the remaining bars in the bar list and extending them using the curve criteria.

In addition to dashed lines, the contest also required the detection of some object classes, including dashed and solid line rectangles, dashed and solid line triangles, dashed line hatched area (rectangle or circle), and the ability of segmenting text from drawings. We fully implemented these requirements in the program.

Empty blocks are detected using a strategy similar to the generic recognition algorithm [10]. The loop comprising the same type of lines is recognized by stepwise finding lines of the same type (i.e., solid lines, dashed lines, or dash-dotted lines, etc.). The dashed lines or dash-dotted lines rather than individual dashes are found as the components of the block. This global method of detection of dashed blocks is much easier than finding dash segments, which encountered difficulties at corners and junctions (i.e., the shapes "T", "+", "L", "X", etc.), because the dashed lines can go through these junctions or reach the corners. The block line type (solid or dashed) can be known from the constituent border lines. The shape of the block is determined from the number of constituent border lines, i.e., it is a triangle if the number is three, and it is a rectangle if the number is four and the neighboring border lines are perpendicular each other.

For hatched area, simultaneous detection of both the hatching dashed lines and the chain of solid border lines are conducted. It is also within the frame of the generic recognition algorithm [10]. After the hatched block is detected, the block shape is determined by the number of border lines and their geometry. If the number is four and the neighboring border lines are perpendicular to each other, it is classified as a dashed line hatched rectangle. If the number is higher, a circle approximation is done to these border lines. If a center and a radius are successfully found, the block is classified as a dashed line hatched circle and the chain of border lines is replaced with the detected circle. The hatched area detection is a topic of another paper [11].

Text segmentation, discussed in detail in [12], is done immediately after vectorization in case of complex drawing. Most of the text is successfully grouped in textboxes. This decreases the false alarm rate of dashed line detection because characters usually have short strokes, which may also be considered as dashes. But it also decreases the recognition rate of dashed curves because dashes are mis-grouped as part of text such as in Figure 7.

5. Experimental Results

Our dashed line detection program was tested with many test drawings during the contest. Figure 6 is a part of noisy, randomly generated drawing used during the contest, where 100% of the dashed lines were correctly detected, with only one false alarm. Table 1, which was generated automatically by the evaluation program, presents evaluation results.

Figure 7 shows two parts of a complex level test image from the contest. All dashed lines were detected, but a flat curve was recognized as straight, and one dash was considered as a part of a textbox, breaking the curve into two parts. Figure 8 is a part of a medium level test image from the contest, where there is a dash-dotted triangle. Our program successfully detected it. Figure 9 is also a part of a complex level test image. Our program successfully detected out a dashed line hatched circle. The black dot in the center is the center of the detected circle. In all images, the black are original pixels and the light lines are the vectors extracted from the raster image.

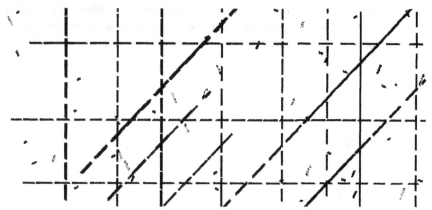

Fig. 6: Results of detection of dashed lines in the contest.

Table 1. Our program's detection rate on the drawing in Figure 6, evaluated by an automated evaluation program at the contest.

```
:::::::::::::::::::::::::::::::::
::    Detection Rate    ::
:::::::::::::::::::::::::::::::::
```

| | Detected Lines | | |
	same as line-type i	Not line-type i	Not detected
Ground truth line-type i	100.00% correct	0.00% mis-label	0.00% mis-detect
Not ground truth	4.55% false-alarm		

6. Conclusion

We have presented the program that won in the dashed line detection contest held during the First International Workshop on Graphics Recognition, August 9-11, 1995. The contest required automatic detection of dashed lines on test drawings at three difficulty levels: simple, medium and complex. The random images contained dashed and dash-dotted lines in straight and curved shapes, and even interwoven texts along with noise. Our program performed very well, and successfully detected the dashed lines (and other objects) in all drawings at all three required complexity levels. The underlying mechanism in the techniques is a sequential stepwise recovery of components that meet certain continuity conditions. Its success encourages us to use its revised version in the detection of discontinuous lines in real life noisy engineering drawings, a problem which is not yet completely solved.

Fig. 7: Results of dashed curves detection and text segmentation. The dashed curve detected at the left would have extended to the right dashed line if one dash had not been included in a textbox. The dashed curve on the right was misdetected as a dashed straight line.

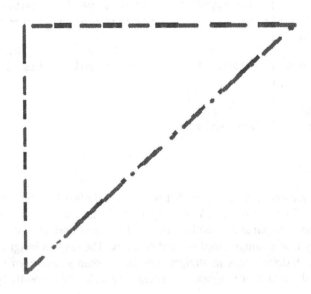

Fig. 8: Results of dash-dotted triangle detection.

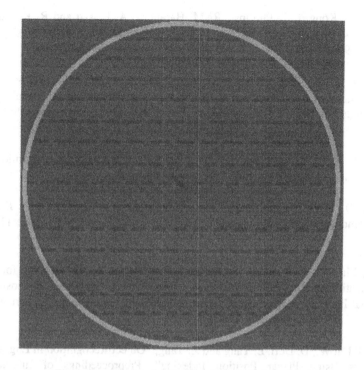

Fig. 9: Results of dashed line hatched circle detection.

References

[1] L. Boatto et al., "An Interpretation System for Land Register Maps", IEEE Computer, 1992, 25(7), 25-32.

[2] P. Vaxiviere, and K. Tombre, "Celesstin: CAD Conversion of Mechanical Drawings", IEEE Computer, 1992, 25(7), 46-54.

[3] R. Kasturi, S. T. Bow, W. El-Masri, J. Shah, J. R. Gattiker, and U. B. Mokate, "A System for Interpretation of Line Drawings", IEEE Transactions on PAMI, 1990, 12(10), 978-992.

[4] S. H. Joseph and T. P. Pridmore, "Knowledge-Directed Interpretation of Mechanical Engineering Drawings", IEEE Trans. on PAMI, 1992, 14(9), 928-940.

[5] B. Kong, I. T. Phillips, R. M. Haralick, A. Prasad and R. Kasturi, "A Benchmark: Performance Evaluation of Dashed-Line Detection Algorithms", In this volume, 1996.

[6] I. Chai and D. Dori, "Orthogonal Zig-Zag: An Efficient Method for Extracting Lines from Engineering Drawings", Visual Form, ed. C. Arcelli, L. P. Cordella and G. Sanniti di Baja, Plenum Press, New York and London, 1992, 127-136.

[7] D. Dori, Y. Liang and I. Chai, "Spare Pixel Recognition of Primitives in Engineering Drawings", Machine Vision and Applications, 1993, 6, 69-82.

[8] D. Dori, "Vector-Based Arc Segmentation in the Machine Drawing Understanding System Environment", IEEE Transactions on PAMI, 1995, 11, 959-971.

[9] Liu W. and D. Dori, "Sparse Pixel Tracking: A New Algorithm for Basic Vectorization of Engineering Drawings", Proc. of the 12th Israel Symposium on Artificial Intelligence, Computer Vision and Neural Networks, 1996, February 4-5, Tel Aviv University, Tel Aviv, Israel.

[10] Liu W., D. Dori, L. Tang and Z. Tang, "Object Recognition in Engineering Drawings Using Planar Position Indexing", Preproceedings of International Workshop on Graphics Recognition, 1995, August 10-11, University Park, PA, USA, 53-61.

[11] D. Dori and Liu W., "Recognition of Hatching Lines in Engineering Drawings", Submitted to Pattern Recognition Letters.

[12] Liu W. and D. Dori, "Vector Based Text Segmentation in Engineering Drawings", Proc. of the 12th Israel Symposium on Artificial Intelligence, Computer Vision and Neural Networks, 1996, February 4-5, Tel Aviv University, Tel Aviv, Israel.

Summary and Recommendations

Rangachar Kasturi and Karl Tombre

The Graphics Recognition Workshop concluded with a session in which panel leaders from various sessions presented a summary of their session's presentations and recommendations for further research. Principal points made at these presentations are summarized below. More detailed reports received from two of the panels are included as appendixes.

1 Summary

- Real-life drawings are large, complex, and noisy. Despite much progress in recent years in graphics recognition, automated conversion of such documents will continue to be a significant research problem in the foreseeable future.
- One of the hard problems in graphics recognition continues to be text-graphics separation, especially when a part of the text is touching graphics. This is very common in most types of drawings and hence warrants urgent attention.
- Large quantities of scanned images of documents are becoming readily available. Advances in hardware and on-board software in scanners has made this possible. Many hybrid raster/vector interactive document data management systems are also already available commercially or are in development. In such systems, semiautomatic conversion is generally accepted as a compromise with simple, routine tasks done by machine. Such systems provide a means for integrating graphics recognition algorithms as they become mature while putting digital images to immediate use, without the need for waiting until all graphics recognition problems are solved. Good human-computer interfaces are critical for the success of such interactive conversion systems.
- In view of the high design and development costs, large volumes of similar documents are a prerequisite for cost-effectiveness. Utility company drawings are good examples where this requirement is easily satisfied.
- In some application domains redundant information is often available in different drawings of the same set. Consistency checks at various stages of conversion using such drawing sets would help to improve the overall recognition accuracy.
- Very little work has been done for performance evaluation of complete systems, although some work has been done in the evaluation of individual modules. Formal methods for system-level evaluation would help to determine the choice of individual modules and their associated parameters so as to maximize system-level performance. Formal measures for calculating system cost/complexity vs. performance, misdetection vs. recognition accuracy, and cost-benefit trade-offs are needed.
- Performance evaluation is often limited to synthetically generated images. Noise and perturbation models are needed to make these images as similar to actual scanned drawings as possible. Synthetic images help in the precise quantification of performance, but testing on real images is still needed to validate such performance data.

- Migration from electronic files of 2-D CAD to 3-D CAD systems has created new opportunities for extending the usefulness of Graphics Recognition algorithms beyond paper drawing conversion. Although such systems benefit from noise-free electronic input data, problems caused by missing or intentionally omitted details in some of the 2-D views of a 3-D object continue to be challenging. Unfortunately, in some instances in which the original company is no longer in business, conversion should be done using paper copies of drawings, since no information on the proprietary data format is available. Naturally, the ultimate goal of graphics recognition is to enable both machines and humans to understand data presented in exactly the same format!
- Graphics recognition is also useful as a human-computer interface in situations where a human would draw sketches and the computer would interpret them. This method is particularly useful for inputting complex parts into CAD systems; it is easier for engineers/architects to draw 2-D orthographic/cross-sectional views rather than input 3-D object models.
- Search and retrieval systems based on graphical content will experience strong demand in digital library projects and image database systems. Graphical search and indexing capabilities would be crucial for the success of such systems. While some of these requirements would require recognition of paper-based symbols, a larger need exists for systems that would retrieve objects which are similar to a sketch input by a user. Retrieval of similar objects is a well-known problem in copyright/patent processing; computer-based systems are essential to search large repositories of documents on file for such comparisons.
- Color and gray-level processing is still not very common in document processing, in general, and graphics recognition, in particular. While color processing would be very valuable to separate different layers of data in a map, it is still not clear if the additional processing necessitated by using color/gray-level in a general document would result in significant improvements in performance.
- Graphics recognition methods are particularly useful for spatial data handling at a cadastral map level. Separation of various layers from a paper map, either by using differences in line patterns or by using color, is a cost-effective procedure for rapid creation of geographic information systems of a local region.
- Methods for representation of knowledge and tools for flexible representation and use of such knowledge are still ad-hoc. A formal engineering approach is needed.

2   Recommendations

- Establish standards to facilitate exchange of data and code among researchers. This would also help in rapid prototyping, using modules developed by other researchers, and improve cooperation among various research groups.
- Create software modules and toolboxes of commonly used and well-known algorithms for graphics recognition to facilitate rapid advancement of research and development. It is well-known that such efforts are highly beneficial to both application developers and researchers. Khoros/Cantata has helped the image processing community tremendously; the Image Understanding Environment is expected to

provide similar facilities for computer vision researchers. We need a similar effort, albeit at a smaller scale, to develop a useful toolbox for all document image analysis application development, in general, and for graphics recognition, in particular.

- Create databases for performance evaluation and meaningful comparison. Develop metrics for reporting performance. Using these databases, conduct systematic experiments comparing characteristics of different algorithms and their break points. Publish papers documenting these results, so that a practicing engineer could choose the algorithms and their parameters which are more appropriate for his/her application, instead of using the traditional trial-and-error method.
- Researchers should continue to address more difficult problems such as segmentation of text touching graphics. Contests similar to the dashed-line detection contest held at this workshop should be organized to encourage innovation.
- Potential improvements in performance by using gray-level and color processing methods should be studied; such methods are of course essential in application domains such as multicolor map processing.

Appendix A: Summary report on Basic Techniques and Symbol-level Recognition

In addition to the overview talk, six papers were presented in this session. The purpose of this summary is to group the talks by topic and to point out novel, cogent, and pervading issues.

This summary has been organized into four sections. The first describes some points made in the overview talk. The second and third sections constitute an attempt to organize the talks by common topic. The second section is titled, Vectorization and Polygonalization, and the third section, Recognition and Learning. The fourth section identifies some common questions made during the session and identifies future work.

Overview Talk — Basic Techniques and Symbol-Level Recognition

Lawrence O'Gorman presented a background and overview of low-level processing techniques as applied to graphics recognition. The main theme of the talk was that preprocessing (that is any processing performed preliminary to obtaining features and in performing recognition) is important to the final recognition results. Very simply, no matter the effectiveness of the recognition stage, if information has been lost or corrupted in previous stages, recognition results will suffer. Though this point may appear to be obvious, many practitioners and researchers in this field continue to neglect the initial stages of processing. A common example of this is the use of binary scanners to enter images. Since the binarization level is fixed, the beginning binary image has much lower quality than if it were to be scanned in gray-scale, then processed using a state-of-the-art adaptive thresholding algorithm.

One point made in this talk was regarding the future of image compression research. This field has reached a point that very little further compression can be gained by the digital signal processing approaches used until now. For binary compression, JBIG is the state-of-the-art standard, and this uses 2-D run-length coding and arithmetic coding.

The compression community is now looking at recognition techniques to make further compression gains. For instance, as a replacement for the current fax standard (CCITT Group 3), researchers are looking at recognizing similar symbol bitmaps on a page. This is a process akin to OCR, but without the final recognition decision. Therefore this is an opportunity for graphics recognition researchers to contribute their work to this important field.

Some families of methods are well-researched and mature in this area of low-level processing, such as thinning and thresholding. There is always room for researchers to make improvements in these methods, but there is a danger when inventing a new method that it is not better than one or more already in the literature.

As evidenced by the papers in this session, there are also areas that are less mature and require more research. Three such areas are critical point detection of lines and curves, recognition of general graphics entities, and learning so as to make the recognition process adaptive to different applications and types of images. Two factors that make these areas difficult are the presence of noise and the need for scale adaptation for different size features in the same or different images.

Vectorization and Polygonalization

Atul Chhabra presented work on vectorization of telephone company drawings, in particular, identification of horizontal lines in these drawings. Since this is a very application-specific task, the objective was to achieve very high recognition results in the presence of a large amount of noise, and to do so quickly. This is an example where efficient and effective results are achieved by using very focused domain knowledge. Important to note is that this is a production system, to be employed in early 1996 by NYNEX Corporation on their thousands of drawings.

Gady Agam presented a morphology-based method to recognize dashed lines. This is another specific application — dashed lines — however, the objective here was to develop a directional morphology method that will adaptively adjust for local line direction. Therefore, the method is very focused, but can be used adaptively for different applications of different dashed line characteristics.

Markus Röösli presented work on high quality vectorization as applied to cadastral maps. Although the application is again very specific, the work employs geometric constraints (as opposed to application-specific constraints) to improve the vectorization results.

Recognition and Learning

Simone Marinai presented a hybrid system for locating low-level graphics items, which was demonstrated for logo recognition. A connectionist model was used here where learning was performed by example. Experimentation was performed with 215 examples and simulated noise. It is evident that work such as this is still in the growing stage (in contrast to more application-specific work such as above); however this goal of learning is ultimately highly desirable.

Dov Dori presented work on knowledge-oriented primitive recognition. Three points are important here. One is that knowledge can be used to enhance the ultimate recognition effectiveness. Secondly, knowledge acquisition is the goal here for truly general and adaptive systems. Finally, iterations of knowledge acquisition improve effectiveness.

Horst Bunke presented work on automatic learning and recognition of graphical symbols. The methodology employed here is attributed relational graphs for the objective of learning new symbols and adding to the symbol database. This is the ultimate objective of our area, that is to recognize currently known objects, and to recognize unknown objects which are added to the lexicon for future recognition as known entities.

Questions and Future Work

One common theme of questions to the speakers was: What is the purpose of the work and how do you measure success? Three answers were volunteered for this:

- Methods with similar purposes should be tested on the same data and their results compared.
- The computational cost is important as well as the effectiveness. For instance, in many instances, it is desirable to sacrifice better results if the methods can run quickly on an inexpensive PC.
- For methods that are employed in industry, there is a simple metric, that is cost-benefit, or money saved or made by using a method.

Another common theme in this session was a question of the tradeoff between application-specificity and generality. The first talk above was very application-specific, however it is demonstrably and economically very effective in real-life application. This is contrasted with less mature algorithms that are still in research mode. The goal of we researchers is generality. However we are in an applied field, so an important question is: in practice, what is the current state of the art in regards to generality versus specificity, and where will the field be in five years?

In regards to future work, what specific areas are important to improve and on which to concentrate our efforts? It was mentioned that, though many low-level methods are mature, it is important to use these in real problems. For this to happen, the methods must be fast and easy to use. (An example of ease of use, is an adaptive thresholding method embedded in a scanner.) The use of recognition techniques for document image compression will be an important future (and current) research area. Recognition for database query use will be an important use of our methods, especially because document and multi-media databases have become more prevalent. Finally, we still have much research to perform for our ultimate objective of adaptively recognizing and learning for new and general document applications.

Summary Report by Lawrence O'Gorman and Keiichi Abe.

Appendix B — Understanding Engineering Drawings and Maps: Research Status and Open Problems

Market needs vs. Research State of the Art

There has been general agreement on the fact that the type of drawings and maps currently handled by most researchers is simple and clean. This is quite different than the quality and complexity of drawings customers wish to process electronically and are having problems with. The quality is sometimes so bad that it is even hard for humans to restore the information conveyed by the paper document. This is true for engineering drawings, maps, utility maps, etc. in which lines are broken, merged, smeared, faded, and the paper is taped, folded, wrinkled, and stained. Paradoxically, however, it is those very documents for which conversion is most needed, because the fact that they are so shabby stems from their frequent and long usage as paper documents, which is a testimony to their importance and popularity.

This raises the question if there is at all any hope for ever achieving the required level of processing and understanding for the real life documents. In response to this, the point has been made that while most important engineering drawings (e.g., in the automotive industry) have already been converted manually by people mostly in countries with low labor cost, drawing understanding is still a viable research area because many companies are now in the process of migrating from first generation 2D CAD systems to modern, 3D volume oriented systems. They need to convert 2D drawings into these new systems. The problem of 3D reconstruction is still open, as are many problems in 2D understanding. The advantage of converting 2D CAD into 3D CAD is that most problems of noise are non-existent here.

Since industry uses a specific CAD system, it expects conversion result to be clean and without any noise, as no CAD system is equipped to deal with contaminated data. Moreover, the interpretation systems are restricted to the hardware of the CAD system if a complete system is to be built. This may further restrict the distribution of good conversion systems once they are available. On the other hand, hardware problems are becoming less prominent with respect to software development costs.

A customary evolving solution is the use of hybrid raster/vector systems that enable editing of both the raster and vector data. Vectorization and conversion is done only at small areas of the drawing where a change needs to be made. The raster is used as a background or a reference and tools are provided for the user. For instance, the map recognition system RoSy (developed by M.O.S.S) works this way. Lines on the map that have been interpreted, can be manipulate by the user, and s/he always has the raster to refer to because the results are overlaid. In a much more primitive version, the user can have the cadastral map as a background, on which he can draw power lines.

Waiting for a brighter future?

Many companies currently scan paper drawings and store them as images, waiting for the technology to further mature and carry out the difficult conversion tasks. A major emerging requirement is the need for graphical indexing, which would enable to carry out graphical queries. The general consensus is that one cannot normally expect a 100%

automation in the process, so human involvement is a necessity. If a system does around 85% or more of the conversion automatically then it is certainly a great contribution.

Interpretation errors and the extent of human intervention

One of the problems is how much of the task should we aim at automating. On one hand, more automation means less human labor, but on the other hand, more errors that may be hard to detect can potentially sneak in as the automation level is increased. If the drawing has an error (i.e., the draftsman made a mistake), then automating the process of recognition will probably mean that this error will go undetected and will be entered into the electronic database. For maps in particular, lacking consistency checks, the error can go undetected. Some companies are so afraid of "contaminating" their electronic database that they are reluctant to use any recognition system. They are very certain about there being errors in the maps and they feel that these will go undetected if a program interprets the maps. Machines may generate errors that are either very obvious and easy to detect, or errors that may go undetected even by highly qualified people. To decrease the likelihood of errors contaminating the documents, a host of consistency checks should be included in any industrial strength system.

The answer to the extent of automation in the interpretation and conversion process is therefore application dependent. For example, in a European project where the conversion of 120,000,000 documents is required, no human intervention is acceptable, so either the input documents must be very clean or a certain amount of undetected errors must be taken into account. In other applications like address reading systems with even a 50% rate of success, human labor is reduced and thus these systems are helpful. The problem lies not in the low recognition rate but in the possibility of the system misdetecting. While in mail handling the only damage may be slowing down the delivery of the wrongly interpreted addresses, in applications like check reading this is of course not tolerable, so the reject rate is set high to avoid misdetection of the largest extent possible.

Considerations similar to those made here have been made in the area of robotics. There too, robots cannot normally replace human labor completely, so we have to contend with a rate of 70-80% replacement. Systems would be most useful if they were able to realize and state what they can do and what they cannot do. However, for most systems, these requirements, at least for now, seem a bit far fetched. The best mix is for the machine to do the routine and easy tasks, leaving to the human operator the decision making task regarding the harder parts of the recognition. To do this effectively, good human machine interfaces must be designed and implemented.

Summary

There is a host of open issues in the graphics recognition research area, and there is no doubt as to the viability and growing importance of the field. Contribution has already been made in specialized areas, for instance, for telephone and power companies which hold huge numbers of drawings with the same syntax and appearance, making the development of a specialized system for processing them cost effective. Image and graphical document databases are two multimedia-enabling technologies that seem to be heavy consumers of graphic based operations such as graphical queries. Likewise, the bi-directionality of exchanging information stored on paper and electronic media will make increasing use of graphic recognition capabilities. Overall, a technology that was originally motivated by the pressing need to convert paper documents containing graphics into electronic formats is becoming more and more useful in a variety of information technology domains never before thought of. Once the technology becomes mature enough, one can envision a team of humans working with robots, where both humans and robots read the same graphical document (displayed on any media) and coordinate their tasks based on understanding the same document.

Summary Report by Dov Dori and Gladys Monagan.

Lecture Notes in Computer Science

For information about Vols. 1–1006

please contact your bookseller or Springer-Verlag